SpringerBriefs in Molecular Science

History of Chemistry

Series Editor

Seth C. Rasmussen, Department of Chemistry and Biochemistry, North Dakota State University, Fargo, ND, USA

Springer Briefs in Molecular Science: History of Chemistry presents concise summaries of historical topics covering all aspects of chemistry, alchemy, and chemical technology. The aim of the series is to provide volumes that are of broad interest to the chemical community, while still retaining a high level of historical scholarship such that they are of interest to both chemists and science historians.

Featuring compact volumes of 50 to 125 pages, the series acts as a venue between articles published in the historical journals and full historical monographs or books.

Typical topics might include:

- An overview or review of an important historical topic of broad interest
- Biographies of prominent scientists, alchemists, or chemical practitioners
- New historical research of interest to the chemical community

Briefs allow authors to present their ideas and readers to absorb them with minimal time investment. Briefs are published as part of Springer's eBook collection, with millions of users worldwide. In addition, Briefs are available for individual print and electronic purchase. Briefs are characterized by fast, global electronic dissemination, standard publishing contracts, easy-to-use manuscript preparation and formatting guidelines, and expedited production schedules. Both solicited and unsolicited manuscripts are considered for publication in this series.

More information about this subseries at http://www.springer.com/series/10127

J. N. Campbell

Bonds That Tie: Chemical Heritage and the Rise of Cannabis Research

 Springer

J. N. Campbell
Independent Scholar
Spring, TX, USA

ISSN 2191-5407 ISSN 2191-5415 (electronic)
SpringerBriefs in Molecular Science
ISSN 2212-991X
History of Chemistry
ISBN 978-3-030-60022-8 ISBN 978-3-030-60023-5 (eBook)
https://doi.org/10.1007/978-3-030-60023-5

This Springer imprint is published by the registered company Springer Nature Switzerland AG
The registered company address is: Gewerbestrasse 11, 6330 Cham, Switzerland

For Jo, with love,
I will know your story...

Acknowledgements

As my brief argues, no research project, chemical or otherwise, is without heritage. Families and friends sacrifice their time that would have been spent with you, so you can work—to my Nik.

An employer understands when you as an independent scholar are reading an article during working hours; thanks to the Big Three for that—Paul, Roel, and Nelson.

Confidants read sentence after sentence and worked tirelessly to assist in the editing process—love to Jo. My son Jackson, a budding historian of the 1980's popular culture continues to inspire me.

I am grateful to all those that helped to make this brief possible, but no more so than Dr. Seth C. Rasmussen. His leadership and support are a constant reminder that professionalism and creativity equate to collegiality. Once again, he trusted a scholar to publish not one, not two, but three briefs. I am grateful.

One major repository of papers that worked assiduously with me was the University of Illinois, where Roger Adams Papers are held. Linda Stahnke Stepp, the Program Officer for Reference at the University of Illinois Archives, and her team, did everything possible to supply me with access to a cache of papers that were seminal to my argument.

Thanks, are also extended to Dr. Allyn Howett of Wake Forest University's School of Medicine. Her research and responses to my queries were essential to the project.

I want to extend my gratitude to Springer and their editorial staff—including Tanja Weyandt, Stephanie Kolb, and Sofia Costa. They serve as a means for "independents," like me, to contribute mightily to this scholarly cause.

Finally, I would like to express my appreciation to Steven M. Rooney, who is a full-time Instructor of Chemistry at Tarrant County Community College in Ft. Worth, Texas. Steve is a good friend and along with his wife Teri, were always close confidants of mine. We lost her back in 2018, but her memory serves as a beacon for both of us, especially in tough times. I appreciate Steve for reviewing multiple drafts of this work and providing valuable renderings for the project. I could not have proceeded without him.

A Note on Spelling

Notice the spelling used of the word *marijuana* is used throughout the brief, instead of the period one—marihuana. The two are considered one and the same. However, the latter spelling was dropped sometime after the Second World War. The original spelling from the nineteenth century was reverted to and became part of the lexicon in the West.

Contents

About the Author

J. N. Campbell is an independent scholar and writer in Houston, Texas. He is the co-author, along with Steven M. Rooney, of the briefs, *How Aspirin Entered Our Medicine Cabinet* (2017), and *A Time-Release History of the Opioid Epidemic* (2018), both published by Springer. He is also a professional turf writer for several publications, covering issues in the world of Thoroughbred racing. After receiving an M.A. in History from the University of Kentucky, he gained a second M.A. from the Parsons School of Design in New York. He lives in Houston, Texas.

Chapter 1
Introduction: Origins of Cannabis Research

This drug is as old as civilization itself. Homer wrote about that it made them forget their homes, and that turned them into swine. In Persia, a thousand years before Christ, there was a religious and military order founded which was called the Assassins and they derived their name from the drug called hashish which is now known in this country as marihuana. They were noted for their acts of cruelty, and the word "assassin" very aptly describes the drug.

—H. J. Anslinger, Commissioner, Bureau of Narcotics, Treasury Department, before the House Ways and Means Committee, 1936 [1].[1]

To sum up, Indian hemp, like many other medicaments, has enjoyed for a time a vogue which is not justified by the results obtained. Therapeutics would not lose much if it were removed from the list of medicaments.

–Dr. J. Bouquet, Hospital Pharmacist, Tunis, quoted by Anslinger from a League of Nations Report [1].

---

[1]Harry Anslinger's version of the past was formed to suit his own needs, and is archetypal misstatement of definitions. Both hashish and marijuana—also called weed, pot or ganja—are parts of the cannabis sativa plant. The major difference between the two is that the term "weed" usually applies to dried pieces of the plant, mainly flower buds, while hash is a paste from resin, or sap of the plant. Hash contains a higher concentration of psychoactive chemicals.

J. N. Campbell, *Bonds That Tie: Chemical Heritage and the Rise of Cannabis Research*, History of Chemistry, https://doi.org/10.1007/978-3-030-60023-5_1

1.1   Bonds of Heritage and Medicine

Hemp is a revolutionary plant; it has heritage.[2] Need clothing or fuel? Hemp provides. Need a drug that can soothe pain? Certain parts of hemp can do it. In the United States, until recently, hemp was seen as a product that had limited capability compared to say, cotton. The federal government only recently moved to set it free. Times are changing for the age-old plant. Currents of thought are now moving in such a direction that the production of what has become known as cannabidiol (aka CBD), for some, has the potential to transform the medical industry—debate is ongoing.[3]

Many countries are embracing hemp's power and its' ability to be a flexible fiber and the progenitor of new biochemical mixtures that are specifically designed to relieve pain. Terms can confuse us. What is the difference between hemp, marijuana, and cannabidiol? All possess chemical properties, but how are they different? What is known is that both hemp and cannabis ultimately come from the same plant; just different parts. Whether its' deemed hemp or cannabis, the distinction depends on a variety of factors (Fig. 1.1). However, despite the fact that the terms hemp and cannabis are often used interchangeably, they do have separate connotations. For instance, a 1976 study published by the International Association of Plant Taxonomy argued that "both hemp and marijuana varieties are of the same genus, cannabis, and the same species, cannabis sativa. There are countless varieties that fall into further classifications within the species cannabis sativa." All of this designation depends on how the plant is grown and utilized. For instance, the term cannabis (or marijuana) is used when describing a cannabis sativa plant that is bred for its' potent, resinous glands (known as trichomes). These contain high amounts of tetrahydrocannabinol (THC), the cannabinoid most known for its psychoactive properties [2].

Conversely, hemp is used to describe a cannabis sativa plant that contains only traces of THC. Hemp is a high-growing plant, typically bred for industrial uses such as oils and topical ointments, as well as fiber for clothing and construction, just to name a few. For instance, in the United States, only products produced from industrial hemp (less than 0.3% THC) are legal to sell, buy, consume, and ship. This single factor (0.3%) is how we can distinguish between what is classified as "hemp" and as opposed to "cannabis." These designations have led to confusion

---

[2]Hemp and its history are a massive topic—spanning centuries. There are some important macro and regional studies on the subject, both popular history and more scholarly. Two to consider are: Doug Fine's, *Hemp Bound: Dispatches from the Front Lines of the Next Agricultural Revolution* (2014); and the relevant regional study by James F. Hopkins, *A History of the Hemp Industry in Kentucky* (1998).

[3]Journals, medical professionals, hacks, soothsayers, and storefronts peddling "CBD" products are rampant throughout the World. Many claim CBD can be used to cure everything from arthritis to COVID-19. Most claims at this point are unsubstantiated. There simply is not enough evidence; yet promise remains.

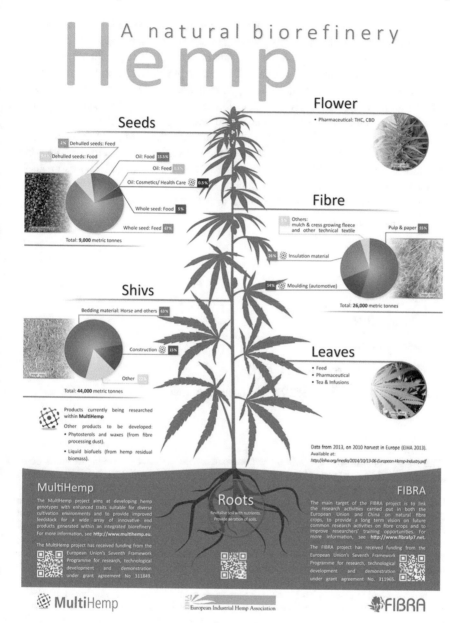

Fig. 1.1 The cycle of hemp (Image courtesy of MultiHemp)

in the U.S. and in other parts of the World that have grappled with legalization [3].[4] To muddy the waters further, many have had trouble distinguishing the difference between cannabis and marijuana. The terms are constantly used interchangeably. Simply put, both refer to the materials of cannabis plants, which in turn lead to the various means of consumption. Smoking "marijuana" is the most common phrase, but drinking, eating, and applying oils or lotions, are also accepted terms related to cannabis consumption throughout the World and in the United States.

The primary psychoactive portion in marijuana is delta 9-tetrahydrocannabinol (THC), which can be found in the resin of the cannabis plant. There are some one hundred known cannabinoids that are unique to their host, and this estimate is growing, with THC being the most famous (Fig. 1.2).

Common types of cannabinoids found in cannabis include:

Cannabichromene (CBC)

Cannabidiol (CBD)

Cannabidolic Acid (CBDA)

Cannabigerol (CBG)

Cannabinol (CBN)

Tetrahydrocannabinol (THC)

Tetrahydrocannabinolic Acid (THCA)

Tetrahydrocannabivarin (THCV), just to name a few.

---

[4]The legality of cannabis for medical and recreational use varies by country, in terms of its possession, distribution, and cultivation, and (pertaining to medical use) how it can be consumed and what medical conditions it supports. These policies in most countries are regulated by three United Nations treaties: the 1961 Single Convention on Narcotic Drugs, the 1971 Convention on Psychotropic Substances, and the 1988 Convention Against Illicit Traffic in Narcotic Drugs and Psychotropic Substances. The use of cannabis for recreational intent is prohibited in most countries. Yet, several nations have adopted policies geared toward decriminalization in order to make singular possession a non-criminal offense (i.e. akin to a minor traffic violation). Elsewhere harsher penalties reside in some Asian and Middle Eastern countries where possession of even small portions can lead to longer-term sentences. Recreational cannabis is legal in Canada, Georgia, South Africa, and Uruguay, plus 11 states, 2 territories, and the District of Columbia in the United States, as well as the Australian Capital Territory in Australia. Legality varies in and subnational jurisdictions pertaining to commercial sale. Limited enforcement exists in places like the Netherlands, where the sale of cannabis is tolerated at licensed coffeeshops. Countries that have legalized medical use of cannabis include: Argentina, Australia, Brazil, Canada, Chile, Colombia, Czech Republic, Ecuador, Germany, Greece, Ireland, Israel, Italy, Jamaica, the Netherlands, New Zealand, Norway, Peru, Poland, Portugal, Switzerland, Thailand, the United Kingdom, and Zambia. In the United States, 33 states and the District of Columbia have legalized the medical use of cannabis, but at the federal level its use remains prohibited for any purpose. To assist with further explanations, see the important article, Thornton J, Nakamura G (1972) The identification of marijuana. J For Sci Soc 12:435–546.

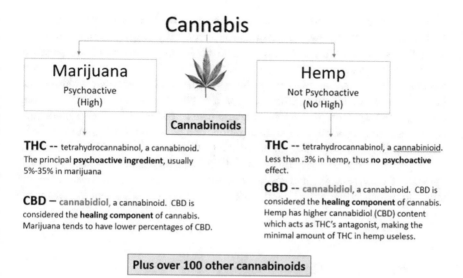

Fig. 1.2 Comparison of Marijuana and Hemp (Image courtesy of APOE 4)

However, cannabidiol (CBD) is becoming more popular as time goes on due to its non-psychoactive qualities [4]. Or more technically correct, phytocannabinoids are the active ingredients found in cannabis (phyto meaning, from a plant). Thus, we could say that there are three working categories of cannabis phenotypes: plants with high levels of THC and low levels of CBD; and vice versa; and finally, hybrid types with equal amounts of each that only form new strains of cannabinoids in this phenotype. In summation, all hemp is cannabis, but not all cannabis is hemp.[5]

As of 2020, medical consumers are focused on one major issue, the effects that drugs have on their health. Over the past century, pharmaceutical companies have come to dominate the market; so much so, that there seems to be a specific drug for everything. Supposedly, progress marches on into a bright future, but not all is certain. Medical chemistry has partnered with biochemistry, which in turn has spawned a new era of biomedicalization by the 1980s [5]. This latest stage is a

---

[5]As will be discussed in Chap. 2, the Marihuana Tax Act of 1937 legitimized the use of the term into the American lexicon by banning its use. Spelled, -irihuana at the time, it was not until after the Second World War that the spelling was changed to the modern -ijuana. Before this period, the substance was associated with trade that was illegal across the southern border of the U.S. Interestingly enough, it was Mexico in the mid-1920s that first banned its trade, not the U.S. Texas did have laws against its purchase, but this was contradicted by the lack of federal legislation. All that changed in 1937. In America, it would not be until the Farm Act 2018, that established a new federal regulatory system under the US Department of Agriculture which aims to facilitate the commercial cultivation, processing, and marketing of hemp. The 2018 Farm Bill removes hemp and its' seeds from the statutory definition of marijuana and the Schedule of Controlled Substances. Hemp is now eligible crop under the federal crop insurance program. The 2018 Bill also allows the transfer of hemp and hemp-derived products across state lines provided it was lawfully produced under a State or Indian Tribal plan or under a license issued under the USDA.

fascinating two-way street. On the one hand you have the pushing by pharmaceutical companies of new products that can specifically target ailments that maybe
a generation before were not conceived to be issues. Conversely, the power of
patients, through the use of self-diagnosis through the Internet, blogs, and Tweets,
has democratized advice. Now more than ever, the market tacks like a ship-of-war
in the Age of Sail. Public opinion can be a powerful motivator which can bring
down a CEO or even a company. Governments now must wait and mark time
before rendering a decision about the legalities of a new drug. It might turn out to be
something they can capitalize on since their budgets are stretched and economic
development is not so easy. In the end, hemp can be the next frontier of development, if only research and testing can move forward to actually produce the
requisite evidence that will allow the public to make decisions about the chemicals
they want to consume.

1.2 Scope of the Current Volume

If we examine the word, "chemical," in the context of post-modern world we inhabit,
it does not always engender confidence. The headline, "A chemical plant exploded
and is fuming," elicits notions of destruction and a major public threat. While "They
died of a chemical overdose," conjures notions of unbridled abuse and addiction.

Speaking medically, we are bombarded with all sorts of "chemical" options that
are packaged and marketed to target a specific malady. Where those chemicals are
created and how they make it to our medicine cabinets continues to be a mix
somewhere between sound chemistry and clever marketing. Laboratories still, as
they have always been, remain the motivating force behind their creation, but they
are not always in control after synthesis takes place. Chemists continue to be the
forgotten portion of a complex nexus perhaps due to their early exit from the
process; meaning marketing and sales ends up driving products more so than the
creators. Thus, we should not forget the impact they have on the future of medicine.

Globally, one of the fastest-growing products that seeks to solve the age-old
problem of pain is cannabidiol, or by its acronym, CBD. Recall that lacking THC,
this drug is attractive due to its non-habit-forming qualities. This volume will
directly deal with its origins and history. Cannabis has a long history, but it was not
until the twentieth century that chemists unlocked the font of potential. Legal
ramifications and the public's reaction to its' use made it a stillborn revolution in
medicinal circles. As I write this brief, CBD's trajectory seems limitless, and yet, it
is full of uncertainty. Despite this paradox, my story begins with hemp, one of the
major forms of cannabis that was and continues to be cultivated across borders.
Growing wild, then manipulated by humans, it has served as a source for all kinds
of products from clothing to oils.

What I am interested in is how did cannabis research lead to CBD?[6] It didn't just spawn on shelves and it was not all that recent. What I found was an intricate process involving the evolution of a chemical heritage. That being, one chemist and then another chemist, and so forth, would move through time taking steps to handoff knowledge, intentionally or not, from one to another. This involved themes like trust, credit, and collaboration, which are integral to courting the very depths and secrets of this branch of science. Bonds and heritage among chemists across national lines and even up against a Cold War, were strong enough in an organic field during the post-World War II era that nothing could halt the testing and investigations into cannabis.[7]

In Part One, my story examines the culture of the early twentieth century laboratory and follows the career of a chemist named Roger Adams. After building a strong chemical program at the University of Illinois, it was Adams who began in the late 1930s to investigate the components of cannabis sativa and was especially interested in the possibilities that it held for unlocking the secrets to pain relief. At this point in medical history both morphine, its' cousin heroin, and other drugs were typecast as villains due to their harsh side effects and addictive qualities. In America these "chemicals" were outlawed at the Federal and state levels. Marijuana was heading in a similar direction. Franklin D. Roosevelt's New Deal politics in the mid-30s and the rising influence of the Federal Bureau of Investigation (FBI) led to the outlawing of its use by the end of the decade. But Adams thought differently. Where others saw drug use, crime, and a means by which to lockup and persecute ethnic minorities, Adams and the chemical heirs of CBD saw opportunity.

In particular, Adams, a chemist who developed an innate sense concerning history, promoted the interplay between chemistry and medicine. Crafting a philosophy while at the University of Illinois that promoted science, he looked to the past just as much as into the future. As this study exhibits, the story of Adams and CBD was only beginning. The Second World War declared the Allies victorious, but it was an American chemistry that sped to the lead of all the sciences, even physics. Chemists became extremely important by not only heading government agencies, but also becoming hugely influential in laboratories that specialized in the making of medicine. Private companies created a "bond" among a university-trained-brethren that would shape policy and spread across the globe as a means of conducting business.

In Part Two, enter heirs to Adams' work on cannabis research. Raphael Mechoulam, an eastern European-born Israeli chemist who, among others, picked up where Adams and his team left off working from the 1960s on. Mechoulam in

---

[6]As of 2020, there are no major investigations of the history of CBD. This is the first to examine the subject from a chemical perspective.

[7]For the impressive piece of scholarship that examines the evolution of technological development see, W. Brian Arthur, *The Nature of Technology: What It Is and How It Evolves* (2009). Among Arthur's insights that serve as inspiration for this brief: (1) technologies inherit capabilities from the technologies that preceded them, (2) components of technology are themselves systems, and (3) all technologies harness and use at least one physical phenomenon.

turn passed the torch on to others who attempted to grapple within a complicated political-economic scenario of drug research, in order to extract cannabinoids from cannabis. By building on a heritage of research, Mechoulam capitalized on one of the most important aspects that chemistry established after 1945—a laboratory beyond nations. With the strengthening of global associations, research flourished. As we will see, some jurisdictions, like the United States, spurned CBD investigations due to their supposed dubious legal status. Despite this, the work of chemists, like Allyn Howlett, led to major cannabinoid breakthroughs in the 1970s and 1980s. In meantime, while these developments proceeded, I will spotlight an anomaly that developed at the University of Mississippi with the opening of the only DEA-sanctioned marijuana testing laboratory in the United States. That entity has made a major impact on whether research and opportunity elsewhere can advance. Finally, the Epilogue will survey the effects of what these chemists chose to develop with cannabidiol as it moved from an American-led laboratory to a global one. This was the literal transformation of CBD along with an already burgeoning biomedical market.

Today, CBD storefronts and a robust online presence are more expansive than ever. The central question though is where is this pointing us? In the end, only chemists hold the key to understanding a complicated process like the one that is associated with cannabinoids. This brief argues that since a chemical heritage created CBD, it can be relied upon again to help us comprehend the future; this is the story of cannabis research as it arrives into the twenty-first century.

References

1. Excerpt of League of Nations document, O.C. 1542 (O) Geneva, February 17, 1937 Advisory committee on traffic in opium and other dangerous drugs, Sub-Committee on cannabis 1 (Report by Dr. J. Bouquet, hospital pharmacist, Tunis, inspector of pharmacies, Tunis, containing answers to questionnaire submitted to the experts) VII, p. 39. http://druglibrary.org/schaffer/hemp/taxact/anslng1.htm. Accessed 2 Jan 2020
2. Pollio A (2016) The name of cannabis: a short guide for nonbotanists. Cannabis and Cannabinoid Res 1(1):234–238. https://doi.org/10.1089/can.2016.0027. Accessed 2 Jan 2020
3. FDA's guidance on "Botanical drug development" available at: https://www.fda.gov/media/93113/download. Accessed 2 Jan 2020
4. Decorte T, Potter G, Bouchard M (eds) (2016) World wide weed: global trends in cannabis cultivation and its control. Routledge, New YorkThis reference should be numbered 4, and the FDA's guidance reference should be returned to 3. Please edit.
5. Clarke A, Mamo L, Fosket JR, Fishman J, Shim J (eds) (2010) Biomedicalization: technoscience, health, and illness in the U.S. Duke University Press, DurhamThis reference should be numbered as 5.

Chapter 2
Part One: The Tying of Early Bonds

2.1 The State of the Early Twentieth Century Laboratory

The first rule of intelligent tinkering is to save all the parts.
—attributed to Aldo Leopold, author and conservationist [1].

Coming at a time when the science of chemistry in this country was entering a new era, he brought to the department a background of culture, a training in investigational science, editorial and executive experience.
—quoted by Roger Adams in his Biographical Memoir of William Albert Noyes, National Academy of Sciences [2].

2.1.1 Cannabis-Related Products: The Case of the "Perfect Food"

Prior to the age when modern industrial laboratories designed complex chemicals geared for humans, and biomedicalization created maladies we never knew existed, there were progenitors that engaged in the making, prescribing, and dissemination of cannabis-related drugs for medical use.[1] Across the globe, grown as hemp, cannabis became a plant with many uses. Complex societies in East and South Asia, in and around the Mediterranean, or in the Atlantic World, perceived that the products of cannabis could be propagated for medical usage. Thus, cannabis was, and still is, one of the most significant drugs ever manipulated by humans.

---

[1]Biomedicalization is a process of technoscientific change that has been occurring since roughly 1985. Its study is concerned with the definition of life. In a post-modern world, medical interventions are designed not just to extend life and promote health, but to enhance our biological selves

© The Author(s), under exclusive license to Springer Nature Switzerland AG 2020
J. N. Campbell, *Bonds That Tie: Chemical Heritage and the Rise of Cannabis Research*, History of Chemistry, https://doi.org/10.1007/978-3-030-60023-5_2

Ancient world physicians and healers practiced forms of materia medica that was the forerunner to twentieth century pharmacology. Outside of the West, some of the most advanced cultures placed a significant premium on understanding that combining cannabis into medicines could actively treat pain and other ailments by utilizing historical plant-based inquires. China, Middle Eastern empires like the Safavids and Ottomans, and Native American tribes engaged in using hemp for medicinal purposes. It was not until the nineteenth century that cannabis was introduced for therapeutic use in Western medicine [3]. One of its early proponents was William Brooke O'Shaughnessy who believed in all-things cannabis. He was an Irish physician well-known for his wide-ranging scientific work in pharmacology, chemistry, and inventions related to telegraphy during the Raj in India. Though officious, garnering many political enemies, his medical research led to the development of intravenous therapy and introduced the therapeutic use of cannabis sativa.[2]

Medically-speaking, cannabis did not stop there. Spreading throughout Europe by 1900, it became part of a milieu of products that were the offspring of the patent medicine trade. These heavily alcohol-backed syrups were loaded with all kinds of sedatives. Made by home chemists in their kitchens, some were even scaled for larger production, while many were infused with a variety of cannabis or coca plant additions [4]. The idea was born fully that you could take everyday food items and mix hemp-related ingredients to form what would become much like CBD-based products is marketed today—a host of healthy products that could be good for you. Industrialization and chemistry bonded to form a revolution in nutrition science.

Proteins built the West, along with sweat, muscle, and ingenuity, is an oft-argued point of view among historians. Whatever the reason or reasons, nutrition became a major business venture for firms looking to take advantage of the new wage earners that were emptying from rural areas and jumping on steamships for faraway lands [5]. Chemically-speaking, beginning with the work of Justus von Liebig during the mid-nineteenth century, and his contest to turn beef into a gelatin that could be preserved and reanimated when added to boiling water, new industrial laboratories across Europe and America sought to capitalize by providing well-packaged items for consumption.[3] One of the most famous remedies dispensed in the home came from a bottle and was carefully poured onto a spoon—castor. Despised by children for its horrid taste, it was billed as an oil that could strengthen constitutions and

---

[2]William Brooke O'Shaughnessy (October 1809–1889) was an Irish physician who performed a wide-range of scientific work in pharmacology, chemistry, and inventions related to telegraphy and its use in India. His medical research led to the development of intravenous therapy and introduced the therapeutic use of Cannabis sativa to Western medicine.

[3]Justus von Liebig (1803–1873) was the son of a pharmaceutical and chemical dealer. He was apprenticed to an apothecary, which before 1820, were loosely linked to the study of chemistry. Dissatisfied under this tutelage he moved to Paris to study under the chemist, Joseph Gay-Lussac. With the help of Alexander von Humboldt, he joined the faculty at the University of Giessen, where he remained until he received the chair in chemistry at the University of Munich in 1852. For an examination of Liebig's beef contest see, Campbell and Rooney, *A Time-Release History of the Opioid Epidemic* (2018), 39–43.

ward off a host of maladies. Firms knew that taste literally mattered; so many worked to do something about health by devising concoctions that could check all the nutritional boxes, and still provide a wonderful gulp of goodness [6].

One of the most telling products of the age that reflected the complex use of many ingredients and sold wildly for over two generations was the Scandinavian-made, Maltos-Cannabis. A Swedish hemp seed-based malt beverage that could be described as a "food remedy" with sedative qualities. Widely-available in Sweden, Denmark, and Norway, it was advertised and reviewed in all sorts of western publications. For instance, an 1899 materia medica guide from Philadelphia explained it as, "a Swedish nutrient in form of a yellowish-white powder, possessing a taste at first saline, later sweetish, and then acrid and bitter." Chemists and physicians separately gave it a ringing endorsement with one even going so far as to describe it as, the "perfect food." Medical journals ranged from a medical appraisal of its soundness to a breakdown of the chemical makeup of this product. This solidified a connection to what was deemed sound nutrition at the time.[4] What helped the cause for selling Maltos-Cannabis were certainly the advertisements of the period. As they had before and would after, print offered veiled attempts to woo potential imbibers into the mix. In this instance, depictions played upon gender roles where a woman-controlled home directed care for her children. As you can see from Fig. 2.1, a mother protects her child by fending off Death with a sickle, and at their side is their greatest asset, Maltos [7]!

More scientific tests (Fig. 2.2) made for a persuasive argument since one of the main components in the mix was not only malt (with the option to have a cacao flavored powder), which possessed high amounts of caloric intake, but also hempseed oil. Thus, Maltos-Cannabis was offering a healthy choice for children and adults alike. Little was mentioned though of its connection to cannabis. This might speak to the idea that those components were widely accepted at the time as being safe for consumers. The hemp seed, depending on what kind was used, could have possessed sedative qualities; but at the time with heroin and cocaine in ready supply, not to mention alcohol-related syrups, cannabis products occupied a fairly innoxious place among the prescriptions of the age. We are not sure of the THC content, though it is possible they could have been significant. By the mid-twentieth century products like Maltos were banned due to their cannabis connections. Going underground, they have only recently resurfaced in the form of cannabidiol products that are sweeping seemingly everywhere. Clearly medical advice and the prescribing of such drugs fights etiolation [8].

---

[4]By 1900, medical journals across the West were focused on professionalizing by publishing articles intended for the trade and for public consumption. They also served as a means for drug companies to indirectly advertise their products. Many would hire physicians to spotlight what they were selling by giving quotes about a pharmaceutical's effectiveness.

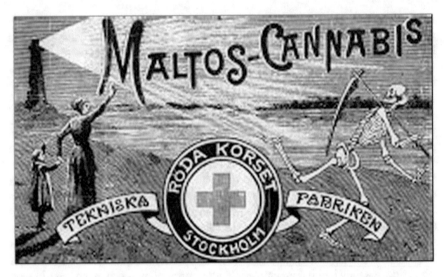

Fig. 2.1 Advertisement (in Swedish) for Maltos-Cannabis (Photo courtesy of the author)

2.1.2   The Chemist Versus the Physician Before 1914

As runaway claims by manufacturers attempted to sell the latest hoax, the relationship between the chemistry laboratory and physician's office before the First World War could not have been more diametrically opposed as far as training. Chemists in Europe spent the better part of the last century establishing themselves as the premier progenitors of scientific medicine, as evidenced by the appraisals made concerning Maltos-Cannabis. Physicians were subordinates and played "catch up," in rudimentary university settings, where training was decentralized, with pharmacists taking the lead when it came to the prescribing and dispensing of drugs. In America, finding a laboratory equipped even with the most rudimentary wares was a desolate business. There were no formally trained biochemists because the discipline did not exist. The effects of the industrialization changed everything for the West, as professionalization took hold and competition led to what, at the time, were revolutionary breakthroughs in biology, medicine, and chemistry [9]. A united Germany took the lead because of their location, and due to the strength behind their science-based university system. Probably most important, they had a high-concentration of companies that discovered an important component that would shape pharmaceuticals to this day—the synthetic dye stuff business.

All this came from the top of the social structure. Designing science and manipulating technology was the province of the few. Loyal Marxists would scoff at this because they believed nothing could occur without a dictatorship of the proletariat, but the rise of corporate capitalism and culture was created by management. Powerful conglomerates scooped up markets, and the race was on to

Fig. 2.2 Chemical
endorsement for
Maltos-Cannabis, 1899
(Photo courtesy of the author)

GENTLEMEN:—The analysis of the
unbroken package (337 grammes) of
"Maltos Cannabis" that we received on
the 14th inst., has to-day been completed by Mr. Kennicott, chemist of
this laboratory, with the following
result:

Fat.	15.01
Salts.	4.75
Albuminoids	7.12
Moisture	6.07
Carbohydrates	67.05
Total	100.00

I find that the substance "Maltos
Cannabis" is excellently prepared, showing an even uniform composition. The
combination of fat, proteid (nitrogen)
and carbohydrate is in such proportions
that the substance contains the essential
food constituents in their most easily
assimilable form.   I consider Maltos
Cannabis a highly nutritious and stimulating food remedy.
Very truly yours,
(Signed)
ADOLPH GEHRMANN, M.D.,
Superintendent Food Inspection.
Municipal Laboratory.

produce products that would solve society's ills and would also meet demand.
Aspirin became the standard bearer for a new age, and it was a powerful story that
combined innovation, competition, and it all happened in a state-of-the-art laboratory. Yet, this was all new territory. Bayer produced it, but they also created
heroin—the shear antithesis of aspirin [10]. At the beginning of the twentieth
century, the issue for companies was that in the past advertising was unseemly, and
the search for profits was not supposed to be at the forefront of business
decision-making. All that changed across the West as print culture accelerated to
new heights. In part, this took advantage of the rise in real wages and leisure time.
Expendable incomes gave a whole new middle class the opportunity to purchase
objects that heretofore were inaccessible.

Purchasing power made pharmaceuticals highly desirable for the public who were suffering under the continued yoke of maladies like whooping cough, cholera, and a host of others. Professional engineering grew rapidly, and the supply was up to the task now that assembly lines and rollouts were packaged, collated, and strategically placed for consumers. But there was a catch. In order for countries to rise to their meteoric potential they needed what Germany developed—streamlining. In the U.S. shortcomings were plenty before 1914. There were no standards of either weights and measures or durability and stress. Proof in medicine had little in form of trials and no means to publish credible results. One of the major roadblocks was the patent system, which remained mainly a service for lone-wolf idea makers tinkering at some makeshift shop in their backyard. The main issue for modern states was how could they make government work for them. Progressive movements, starting with city fathers, began to institute major changes as overcrowding and living conditions deteriorated. Factories were full, but water, sewers, and other public works needed major overhauling. The corporate order needed to fall in line by installing its own manufacturing procedures that would protect consumers. The 1906 Pure Food and Drug Act in the United States is a case in point, along with the creation of the Bureau of Chemistry [11]. One would set standards for consumables, while the other would test them. Neither worked as intended, but the idea was implanted that capital could lead to government enforcement in the service of keeping the public safe.

Across the West, something much more radical was the efforts to change the patent system. Corporations began to hire Jaggers-like attorneys and sue in open court in order to mark their own territory for expansion. No longer would small-timers be able to control the high ground. Filing for a patent became commonplace and a new age was born as giant companies marked their territory by harassing and roadblocking even the smallest impediment. Global competition was moving at a furious pace and in order to keep up, corporate capital had to be fluid and moving in a direction that would assist those that were engaging in the scooping up of raw materials and shipping finished products. Anything and everything were available through mail-order and a catalog system which could be shipped anywhere in the world, pre e-commerce. The drug trade, whether legal or not, was now being backed by new powerful chemistry laboratories, the age of big pharmaceutical conglomerates was born [12].

The medical community in particular greatly benefitted from this rise by corporate drug production because they had undergone their own revolution as 1900 approached. No longer would the "country doctor" who learned by trial and error in the field occupy the main stage. Now a more professionalized set of standards were sweeping the profession. The American Medical Association and numerous other regional organizations were seeking new training and skill developments like never before. Technology, utilized by those large companies, was trickling down to university systems who began to invest in programs that would train the next generation of medical personnel. Physicians, employed in newly designed hospitals, were becoming partners with pharmacies and drug companies, because they helped to disseminate information and sell products [13]. The age of nostrums, alcohol-filled excuses for remedies, were not dead, but the age of aspirin and designed drugs supplanted their power. The search for the "magic bullet," a drug that would specifically target disease and pain, was on. Thus, the chemist, a highly skilled member of a pharmaceutical company, now took the stage.

2.1.3 Experiment: The Hybrid Chemist

If you could engineer a clone of the complete organic chemist what would you want them to be? Would you want her to have a medical degree? Or how about if he had about over 10,000 h of laboratory time? Would being masters of synthesis and extraction be high on the list? Chemists of this nature were few and far between around 1910. Women for the record were barred from such professional endeavors since the movement was, and for that matter still is, fighting for equal pay in exchange for equal work. Programs for training were non-existent. However, German men had opportunity. There a nexus of science mixed with medicine led chemists toward a new horizon that would influence the next generation of what would be called biochemists.

One of the most influential and significant of these early hybrid chemists was Prussian-born and Leipzig-trained Paul Ehrlich.[5] A Noble winner, he became known as the creator of a new derivative from an arsenic compound. Code-named Compound 606, it was effective against malaria infection. In 1905, Fritz Schaudinn and Erich Hoffmann identified a spirochaete bacterium (Treponema pallidum) as the causative organism of syphilis. Ehrlich picked up the significance of this potential breakthrough and tested Compound 606 (chemically arsphenamine) on a syphilis-infected rabbit. Though he did not recognize its effectiveness, a Japanese-trained chemist named Sahachiro Hata joined Ehrlich in Frankfurt after discovering the curative abilities of 606 [14].[6] The German chemist did not feel outdone by a rival from the Pacific. Rather, their two nations were strongly tied by science and a stratified system of heritage. Both saw each other as possessing talents that could help the other. No parts would be pitched; rather what was needed was a combined effort to flesh out potential.

---

[5]Paul Ehrlich (1854–1915) made numerous important contributions to medicine and chemistry in a career that multiple decades. He won the Noble Prize for Physiology and Medicine in 1908 for his combined work with the Russian immunologist, Ilya Mechnikov. Ehrlich's work in bacteriology and experimental pharmacology were based in establishing a framework for a modern research system within the laboratory. He is probably best known for the "magic bullet concept" for the synthesis of antibacterial substances and the development of chemotherapy. His understanding of drug-making would lead to theories associated with the development of antibiotics and the modern-day notion of the more drugs utilized in the proper combinations, the better.

[6]Sahachirō Hata (秦 佐八郎, Hata Sahachirō, 1873–1938) was a prominent Japanese bacteriologist who assisted in developing the Arsphenamine drug in 1909 in the laboratory of Paul Ehrlich. He was nominated for the Nobel Prize in Chemistry in 1911 and for the Nobel Prize in Physiology or Medicine in 1912 and 1913.

Both men, before this fusion, were battling against a routine set of treatments for syphilis that lasted an excruciating 2–4 years. In that time, a patient faced round after round of injections at the hands of a merciless mercury. This left them sterile and their constitutions physically-wrecked. Ehrlich and his team envisioned a new approach to testing their revolutionary drug, through something that would become known as the clinical trial. New processes such as these would become directly spawned from the work in the laboratory, and in this case, it led to the formal naming of one of the world's first target specific drugs—Salvarsan or "saving arsenic." Commercially introduced in 1910, and in 1913 with a less toxic form dubbed Neosalvarsan (Compound 914), the market boomed like powerful cannons shortly before the First World War [15]. These drugs became the principal treatments of syphilis in an age before antibiotics.

Ehrlich's team research on what would become known as the "magic bullet" shook the foundations of pharmaceutical research. The idea that a treatment could target a specific ailment and be engineered in a laboratory, forecast what was to come during the rest of the century and beyond. Yet, even though the door was opened for chemists to incorporate medical fields into their research efforts, it was stillborn revolution [16]. After 1900, especially in the United States, the opportunity for chemistry departments to draw biologists into their fold did not come to fruition. Medical schools became the life boats for some brilliant would be chemists who decided to look elsewhere. There were some exceptions. For instance, the flagship state university in Champaign-Urbana, Illinois would make a home for biochemistry as a subdiscipline under the wings of inorganic and organic chemistry during the 1920s. More on their story later.

The question remains, despite so much promise, why did chemistry turn its back on what would become such an integral part of pharmaceutical manufacturing and research? Everything from aspirin advertising in the 1950s to the addictive effects of opioids in the 1990s to the production and legalization of CBD in 2010s would hinge on who controlled the high ground of biological learning. This was one of those fulcrum moments in science, where a decision or two can make an unbelievable impression on the trajectory of a discipline. With such a moment in play, it all came down to the political turf war that was being waged by the American Chemical Society (ACS), and centralization was the primary point of discussion. To pose it another way, what kind of chemistry would the world have?

2.1.4   What the ACS Did to Biochemistry

If the last quarter of the nineteenth century was an age of unbridled yet targeted industrial expansion from the top, then the next span of the same length, saw governments across the globe attempting to control and modernize in the spirit of the human condition. Progressives believed that science and the shear quantification of fields like chemistry could lead to more efficiency. This would in turn, accelerate everyday life. Governmental programs were booming. The Executive Branch led by

President Theodore Roosevelt, his handpicked successor Taft, and the political science professor, Wilson, saw a new vision for what "federal" could do for people. Civil service and a host of diverse departments would count and recount in order to funnel tax dollars down the food chain. Bureaucracy, lined with red tape, gave teeth to the Department of Agriculture, the newly created Food and Drug Administration, and a host of other entities which would activate throughout this period [11]. Before America's entry into the First World War in 1917, they followed the lead of other Western nations who sought to overturn the past by investing in what they believed was a bright future. Chemistry would be the key.

Recruitment of chemists did not stop with governments at the city to national levels. They were not the only employers that were seeking laboratory personnel. Industrial settings, hospitals, sanitary commissions, and anywhere and everywhere that used, tested, or converted raw materials to finished products were recruiting. Thus, professionalization of this sort naturally led to shoring up the foundations of two major early twentieth century chemical constructs. The first was the advent of the journal [17]. Now of course those had existed for some time, but now at an unprecedented rate they were expanding as new subfields were founded. Second, the professional society, which represented the interests of its members, a sort of union, realized the potential for expansion across disciplines. In order for professional chemists to share their knowledge and exercise their craft, these associations provided valuable outlets for the transference of heritage.

The most centralized and powerful of these associations was the American Chemical Society, which was based in New York City after its founding in 1876. Several of its old titans, such as Ira Remsen, Edgar Fahs Smith, and J.W. Mallet, by 1900 saw the splintering of chemistry as the death knell for the field.[7] Perhaps it was the German influence whispering in their ears, but whatever the reasons they were forming a party line—back the broad-based knowledge. Their worries stemmed from the fact that if hundreds of specialties formed under the umbrella of chemistry, then their ability to govern the field would be compromised. The goal was to have as wide an expanse of members as possible, but they balanced this view with the prospect of keeping the smaller splinter groups from carving away too many layers.

Not everyone in the ACS agreed with this assessment. More moderate perspectives were present. These voices saw the opportunity to include subfields of the

---

[7]Thirty-five chemists met at the College of Pharmacy of the City of New York on April 6, 1876—founding the American Chemical Society. Their first president, John William Draper, delivered his inaugural address at Chickering Hall in New York. From the start, the ACS was committed to sharing its professional pursuits with a public audience. ACS began publishing its flagship journal, the *Journal of the American Chemical Society* (JACS), April 1879. Abstracts, which had appeared in *JACS* since 1897, were given their own publication, *Chemical Abstracts* in January 1907. By 1930, ACS had 18,206 members, 83 local sections and 17 disciplinary divisions. On August 25, 1937, President Franklin D. Roosevelt signed Public Act No. 358, incorporating the society under federal charter. ACS celebrated its centennial year in 1976, at two national meetings with over 10,000 attendees at each. See for more information: https://www.acs.org/content/acs/en/about/history.html.

applied sciences like biochemistry as a boon. They reasoned that it was going to happen anyway; so why not gain their support by reciprocating and bringing these groups into the fold. Institutional growth would benefit from diversity and would allow the ACS to sit on the cutting edge of research. Only cooperation among specialties would take the burden off the universal chemist, and it might lead to breakthroughs that were before unrealized. In the end, there were some half-hearted attempts to keep those that would establish their own subspecialist journals and associations in check. One was *Chemical Abstracts*, a publication that would attempt to attract a broad readership. While it lasted, the ability of the ACS to hold back the flood waters was unsuccessful [18]. Time was not on their side.

One of the reasons this occurred is that the world of chemistry was changing, even before the advent of the First World Wat in 1914. Industrial pharmaceutical outlets, especially in the United States, grew exponentially. Shortly before the world blew up, scientific subfields became primary ones, as with the founding of what would become cornerstone in fields like biochemistry (the *Journal of Biological Chemistry* comes to mind). A field such as this began to gravitate less to mainline chemistry departments at major universities, and instead found a home into the welcome arms of medical schools. After 1920, biochemical studies became a fixture in the curriculum of major medical programs [19].

Chemistry would have to wait and bide its time before it was strong enough to woo the biologists in lab coats back to the fold. In the meantime, there were important questions concerning the developments of pharmaceuticals and pain relief medicine that would not be answered by the science professionals who might know best. Since health care providers and chemists did not see eye to eye when it came to their commonalities, some drugs that could have assisted patients fell into the hands of those that had no scientific background in understanding how to treat pain. A case in point that we could test was the story of marijuana and its uses during the 1930s. That is when a chemist at the University of Illinois began to investigate the power of this age-old drug. His name was Roger Adams.

2.2 Chemical Encouragement According to Roger Adams

The chemist needs encouragement and with this stimulation results may be expected.

—Roger Adams, University of Illinois, unpublished manuscript, What the Chemist Has Done for the Physician, c.1932 [20]

2.2.1 A New Kind of American Chemist Emerges

It could be argued rather persuasively that Roger Adams was an American chemist born into a world at just the right moment. His life, from 1889 to 1971 was reflective of the changes ahead.[8] Slowly, and after decades of incremental growth, the United States emerged from the First World War a different country. European nations went all in, and for the most part, all lost. Some thought they won (i.e. Britain and France), while others bowed out early (i.e. Russia). Financiers and industries scored major victories of course, and technologies brought modernism center stage. Much has been made of the beginnings of an American century; it seemed like everything after 1918 was moving for the Republic in a direction of upward economic, political, and social mobility.

Now, the United States could officially be considered a world power; prerequisites checked off, a navy, colonies, and exports from culture to imperialism. Their brand of science also was making waves as industrial capacities continued, even though Progressivism into the 1920s did not. Business boomed and of course though it would eventually all come crashing down in a global Great Depression, the advancements were fixed. War prosecuted meant economies could be scaled to match any desire, and that is exactly what members of the chemical community did —they soared.

What before was a wild west of chemical companies squabbling over contracts to make the best nitrogen-based gasses that would blister the lungs of the enemy, now normalcy (a crude word from the age) returned to stabilize the industry. Pharmaceutical companies flourished during this period of deregulation and invested huge amounts of capital in laboratories that would become conveyors for a host of new products [21]. This led to an era of increased recruitment and those with science-based training or technical know-how had the opportunity to learn by doing. Since universities were coming online and upping the ante when it came to their organic curriculum, it seemed that thriving times were ahead.

Harvard University was one of the locusts for chemical activity and its department attracted young minds that were full of zest and fervor. The biomedical program was in its beginning stages as a new hospital structure was coming to fruition. Leading their chemistry department was a future Noble Prize winner, Theodore William Richards, who won the award in 1914 for his accurate determinations of the atomic weight of a large number of chemical elements. One of his students in 1905 was a sixteen-year-old kid who showed academic rigor and much promise—Roger Adams. Growing up in nearby Boston, Adams inserted himself into a chemical heritage that was accelerating. Graduating in four years, he went directly into graduate school at Harvard, and then took the path that most chemical

---

[8]Roger Adams (1889–1971) was an American organic chemist known for the eponymous Adams' catalyst, and his work did much to determine the composition of naturally occurring substances such as complex vegetable oils and plant alkaloids. As the Department Chair at the University of Illinois's Chemistry Department from 1926 to 1954, he also greatly influenced graduate education in America, taught over 250 Ph.D. students and postgraduate students. Adams also served as a scientist at the highest levels of American government during the First and Second World Wars. For comprehensive access to his work; see, the *Roger Adams Papers*, 1812–1971, housed at the University of Illinois Archives, https://archon.library.illinois.edu/?p=collections/controlcard&id=3741.

students did, he traveled to study in Germany in 1913. This was clearly a seminal event. What he had the chance to observe was cutting edge work being done at the Kaiser Wilhelm Institute in Dahlem under the discerning eye of Richard M. Willstätter; who, incidentally, won the Nobel the year after Richards. Had Adams not travelled to Central Europe when he did, he might have missed the experience of pre-war Germany under the Kaiser, and its powerful commitment to this science. For nearly two centuries, German states had innovated and outdistanced themselves from every other competitor. Fueled by chemical developments it made their economy robust and in 1870 a unified Germany made quick work of overtaking Britain when it came to industrial capacities [22].

Adams got a full tour. He returned after a year, energized and ready to take a post at Harvard, albeit temporarily. That set his career in motion when he received a call from William A. Noyes at the University of Illinois. State universities were relics from the expansion period before and after the Civil War, but they offered something Ivy schools did not: a blank canvas. Adams arrived just as America's entry into the Great War started, and with his connections, he quickly made his way to Washington to take part in the planning stages and production scaling of the war effort. Food, chemicals, munitions, and everything else was in short supply since Europe was cutoff during the conflict, so the United States had to improvise. Synthetics, the backbone of the chemical industry, had to be made from scratch and laboratories were now moving at breakneck speed. Adams got a front row seat for innovations that were only dreamed of in the previous century [16]. By 1919, as the supposed winners met in Paris to redesign lines on maps from Eastern Europe to the Pacific, Adams returned to ensconce himself in the Illinois Chemistry Department. It would be a long stay.

2.2.2   Chief Illini: The House that Adams Built

In 1921, Percy May saw the printing of his 3rd edition of *The Chemistry of Synthetic Drugs*. Sounds momentous? May must have been proud since most would admit that multiple printings equaled success. We wouldn't know whether he saw the review published in the *Journal of the American Chemical Society*, but if he had, which scholars are prone to do, he would have exhibited a mix of emotions. The reviewer was a young professor of chemistry at the University of Illinois, Roger Adams, and he offered a measured rebuke of the work by calling it a virtual identical version to the previous edition. Though Adams did not care much for the attempt, he took the opportunity to review several drugs that were released over the past three years since the end of the Great War. His technical expertise and attention to detail cajoled the reader into looking elsewhere, while at the same time he admitted as deftly as possible that the volume still possessed value. That was Adams, a chemist who could offer suggestions without criticizing an author for a book he did not write [23].

Fig. 2.3 "The Chief," Roger Adams, 1935 (Photo courtesy of Edgar Fahs Smith Collection, Kislak Center, University of Pennsylvania)

Roger Adams 1935

That sense of heritage and suggestive tone that he briefly exuded in that one-page review encapsulated what Adams was to build at Illinois (Fig. 2.3). He was particularly interested in graduate education and having his charges invest in research projects that were individually structured to their needs. This meant giving them some confidence early on by asking them to publish something on their own as soon as they could. Adams was affectionately and respectfully called the "Chief," probably an allusion to the school's mascot who as a Native American would ride a horse into the football stadium someday [24]. Nevertheless, Adams was all about inspiration and he mastered the art of extroversions. When students witnessed this behavior, they were more likely to emulate it than reject it. Mentors have a way of exercising strong influences, for better or worse, on those under them. Adams knew how to get the very best from those around him, including his own colleagues.

Probably most significant was Adams commitment to building a national program with coverage across chemical subjects. Once he became Chair in 1926, he headed what at the time was a unique design to a department. Other schools, such as

MIT and Berkeley, had strengths in physical chemistry. The Illinois faculty became deeper in teachable subjects, and Adams understood that a commitment to subfields like biochemistry, would make them not only the premier state-sponsored school, but one of the best in the world. Adams would prove the subdiscipline could work in the most rapid means available. The department turned out students who could work in every chemical sector imaginable. The key to this was a balanced approach to education and research. Teaching undergraduates was important, but Illinois also invested heavily in new research avenues. Adams saw firsthand during his stay in Germany how individual chemists could harbor too much authoritarian control. He sought to democratize chemical studies by spreading grants across organic and inorganic studies, instead of pitting mentors against one another [19]. Broad-based planning and pulling together as he often said, laid the foundation for a long-lasting approach to teaching and research. Adams was not against going into battle with sword and shield in hand. Whether it was challenging deans over teaching schedules or writing letters to possible donors for funding, he was willing to do what was necessary to secure futures.

He did not stop there. Adams envisioned from the start that lessons from World War I had far-reaching effects on chemistry. He had heard about students during the summers working on research projects in "preps" labs that had successfully transferred their operations to companies who then invested in product lines. Adams saw the germ of something special. He realized that publishing this data in a regular volume could serve a dual purpose. On the one hand, it allowed graduate students to have their efforts put into what would become the iconic red and green bound volumes; the other, was turning research into viable capital that could fund future projects for the Department. Dubbed *Organic Syntheses*, Adams made it circulated among colleagues and internationally. A companion series, *Organic Reactions*, followed during the 1940s; was based on the same concept of exporting knowledge and expanding fields [25].

One of Adams' lasting legacies was the establishment of a chemical heritage. His influence on the field was reflected in the different types of students that graduated from its hallways. These students carried the torch and held major positions in other universities, privately-held companies or in government. The stats were staggering. Adams helped to train 3% of the Ph.D.'s in chemistry in the United States during the 1920s, and that output increased before the start of the Second World War. Few university-placed chemists during his tenure in Champaign-Urbana could rival his influence. His commitment, drive, and determination secured his place in the pantheon of those that ascribed to the term "greatness" [26]. The best was yet to come as both the international community, and the US federal government courted Adams' expertise and his approach to chemistry. A catastrophic conflict would bring new opportunities for the house that Adams built.[9]

---

[9]Adams was also extremely interested in policymaking and the economics of chemistry. In several published works, he argues that the high cost of medical care will have an adverse effect on

2.2.3 A 1936 Conference Call

Since the mid-nineteenth century, the international community that practiced chemistry developed a collaborative approach to their craft. At times, they transcended borders and squabbles to deliver on the promise to share internationally. Pride in the imagined communities known as nations was not totally excluded, mind you, because there were times when Russians, Germans, French, British, and other European nations flexed their discoveries to mark territory. A Nobel Prize and other international awards meant something, but so did chemical secrets for making war. As mentioned, Americans in 1900 upped their game and joined the community of chemists. They added their own organic solutions to the mix, but it was Germany's defeat in the First World War that really set them on a trajectory for success. As former German assets (most famously Bayer) were seized by wings of the US government, a new age began for science which would have a profound impact on chemistry once the Second World War was concluded.

For Roger Adams, international collaborations were de rigueur and an excellent opportunity for members in the field to learn to think differently. Adams was mindful of his place within the milieu of his subject. To put it another way, he understood that heritage and legacies flowed through both his students and his colleagues [27]. By the 1930s, he had helped to create a dominant department that was committed to turning out a variety of different types of chemists who were ready for real-world experience. During this timeframe, he had a wife and family, served as the head of the ACS, and the University of Illinois wanted to offer him the Presidency. To boot, other universities constantly courted his services, and wooed him with titles that would make others swoon. Adams was unmoved though. He had a good gig, and the makings of a lasting legacy. But, after all he had done, he still needed to stretch his legs abroad and soak in the international scene.

The year 1936 dawned with many political dreams, but mostly the continued awakening of nightmares. The Democrats in America hoped for the re-election of their own version of the pied-piper, FDR. An adherent of Keynesianism, he was a politician who led a spirited attack on the Great Depression with limited short-term economic success. Across the Atlantic, dictators sold their wares to the proletariat in the shape of fear. Stalin, Mussolini, Franco, and of course, Hitler, all wanted their own place in the pantheon of benevolent leaders. All fell woefully short. The Fuhrer

---

society-at-large. He uses a study in 1925 by Dr. Ray Lyman Wilbur, former president of Stanford University, who at the time served as Secretary of Commerce. The committee found the average annual earnings for 30 million people was around $2000.00 per family. Total expenses for a household of four was estimated to be around $1550.00 per year. With little left over for medical expenses, people would simply have to go to a free clinic or do without care. Of 11,000 persons disabled by illness, Met Life Insurance found 35% were not attended by physicians. The problem of rising medical costs appears to have been a problem even in the early twentieth century as it is today. Adams seemed convinced that drugs should be no higher than the cost of any other chemicals. His prediction would ring true that the only option for families inhibited by cash flow would be less expensive chemicals—what we now call, generic drugs.

experienced a major defeat when some of his best Aryan athletes went down in flames against some especially tough competition in the form of Jesse Owens, an American track star who was an African American. Storm clouds were on the horizon as each of these dictators would enact a terrible price on not only their enemies, but their own people. Governments were moving in a whole new direction, and it was going to involve a massive shift in political interests.

It was in this soupy climate, the same year (1936) that Roger Adams took what could be perceived as a jaunt across the Continent—a European expedition. Like many Americans, he wanted to take in local culture, but he was also very keen on making chemical connections with colleagues that he previously contacted by post. Like conference attending today, the opportunity presented itself to experience local cultures and interactions. Serving as a delegate to the 12th International Conference on Chemistry meant something to Adams. He knew that calling on colleagues in the field and probing their methodologies in situ could lead to valuable ideas for projects for his students back home [24]. Afterall, international opportunities afforded the exchange of chemical heritages. Conferences were now commonplace, and they were excellent opportunities to renew old kinship ties and make new ones.

One particular set of meetings were extremely instructive to Adams. After stops in beleaguered Britain and a bestriding Germany, he went to progressive Zurich, Switzerland to meet with a former friend and colleague, Leopold Ruzicka [24].[10] As one of the imminent global chemists, Ruzicka commanded a cadre of graduate students and experienced postdoctoral fellows. His work on large ring compounds and other natural products further stirred Adams' thoughts about the growing influence of pharmaceuticals. Always interested in new avenues, as we shall see, he had growing concerns about the power of drug-making and its influence on human beings and their health.

While he was in Zurich, he also met with two other scientists who were working on a myriad of projects that engaged personal health. Paul Karrer, attached to the major university in town, filled Adams in on his research concerning vitamins and carotenoids, which were the building blocks that won him part of the 1937 Nobel Prize. Like Adams, Karrer had a long career and his voluminous dossier of papers published reflected his active mind.[11] Along with lunching with the father of lactoflavin, Adams had a thought-provoking meeting with an old chum from the Dahlem days, Arthur Stoll, who now served as the vice president of the

---

[10]Leopold Ružička (1887–1976) was a Croatian-Swiss scientist and joint winner of the 1939 Nobel Prize in Chemistry. Working most of his life in Switzerland, he received eight doctorates in science, medicine, and law; seven prizes and medals; and twenty-four honorary memberships in chemical, biochemical, and other scientific societies.

[11]Paul Karrer (1889–1971) was born in Moscow, Russia to Swiss nationals. His family returned to Switzerland where later he studied chemistry at the University of Zurich under Alfred Werner. After gaining his Ph.D. in 1911, he spent a further year as assistant in the Chemical Institute. He then took a post as chemist with Paul Ehrlich at the Georg Speyer Haus, Frankfurt-am-Main. In 1919 he became Professor of Chemistry and Director of the Chemical Institute. Karrer published papers, and received many honors and awards, including the Nobel Prize in 1937.

pharmaceutical company Sandoz. At this point, Stoll had already devoted his service to the company for almost two decades. Both he and Karrer, Adams thought, were perfect examples of what a modern chemist could become. They were two halves of a profession that serviced the student and the public [24]. Their example acted as a wellspring for Adams to hope that his own career could balance both worlds.

Despite this constant contact that Adams experienced on his European trip in 1936, he had a life-altering experience that would change the trajectory of his work forever. Back in the 1920s one of his brightest students was a kid from Iowa who had a knack for chemistry. He wanted to study English but gravitated to the field of chemistry after a mentor suggested he apply his aptitude to the sciences. His brilliant mind was tempered and at times paralyzed by depression and thoughts of suicide. The last time Adams saw Wallace Carothers was on the side of a mountain of the Swiss Alps. He would be dead within a year. Adams mourned the loss of such a promising chemist, and friend.

2.2.4 When Chemical Encouragement Failed

Before Wallace Carothers left this world, he had a bright future that was already coming to fruition in the mid-1920s. His career trajectory was meteoric, and any mentor named the "Chief" or otherwise would have beamed with pride at his accolades. His rise led him in 1921 to obtain a MA from the University of Illinois under Carl Marvel. After a brief stint in jackrabbit country at the University of South Dakota, Carothers returned to work on his Ph.D. under Adams. That choice set him on a path enlivened by chemical heritage. Adams was heading towards being fully ensconced in his Illini chiefdom, and several key students were on hand to witness his ascendancy. Carothers was one of them, and he knew Adams would become a lifelong colleague. Adams' method seemed to work to perfection [28].[12]

By 1924, Carothers earned his doctorate in chemistry, and headed to a position at Harvard University in what he thought would be a feather in his cap. There the venerable James B. Conant, a future president of the esteemed university, marveled at Carothers' ability to harness interpretation. Conant would not have Carothers in his midst for long. Research became an overwhelming draw for the young chemist,

---

[12]Wallace Hume Carothers (1896–1937) was an American chemist, inventor, and the leader of organic chemistry at DuPont. Though he is credited with the invention of nylon, Carothers was also the group leader at the DuPont Experimental Station laboratory, near Wilmington, Delaware —the base of polymer research. Carothers also was instrumental in the development of neoprene. After receiving his Ph.D., he taught at several universities before he was hired by DuPont to work on research. He married Helen Sweetman on February 21, 1936. Carothers was troubled by periods of depression and was exacerbated by the death of his sister. On April 28, 1937 he committed suicide by drinking potassium cyanide. He had a daughter, Jane, who was born on November 27, 1937.

and he made a fateful decision to take his skills to the world of company chemistry —Dupont. For those of you that do not know, Dupont was one of the chemical giants of the twentieth century. Its founder, H.F. du Pont, was considered one of the titans of the industry. Joining their team meant a move to laboratories that were fully stocked with the latest technologies.

Carothers arrived in sunny Wilmington, Delaware hesitant, and although he possessed some of the latest technical knowledge concerning liquid polymers, his psychological state was anything but sound. In fact, he was so against coming to DuPont he even explained to them how he suffered from "neurotic spells of diminished capacity" [24]. Dupont was undeterred. They doubled his salary and with the lower cost of living, a more rural setting might be good for Carothers. Adams knew of his struggles having worked with him closely.[13] Laboratory activity required diligence, but it can also be an intimate and a focused affair for individuals. One is laboring in tight spaces, so it becomes incumbent upon groups of chemists to get along.

There is a very interesting 1931 photograph (Fig. 2.4) of Carothers holding a piece of his newly invented neoprene, another of the products he helped to create. Of course, it was also nylon that was a miracle fabric that was lightweight and pliable. In the photograph, there is the chemist, with both hands on either end, showing how far it can go. His round spectacles seem to gleam, and he possesses a wry expression that connotes the phrase, "look what this can do." By the time of that image of Carothers there were all kinds of signposts that he was drowning in sorrow and despair. He would have these episodes where the pain and suffering were so intense, he would go missing for days and weeks on end. Once he ended up in a nearby Baltimore clinic that specialized in psychiatry. After that stint, he temporarily found his feet and returned to DuPont. Once back in his element he experimented with what would become known famously as nylon. By leading a group of chemists that produced a half-ounce of the stuff from hexamethylenedi-amine and adipic acid, they created a polyamide 6-6, which was part of the greater family of nylon [28].

Despite all the accolades that came from his contributions to chemistry (he was the first industrial organic chemist to be inducted into the National Academy of Sciences), it was not enough to save Carothers. His bouts with what later genera-tions would come to understand as a disease, would not abate. We are not sure what kind of psychoactive drugs he was given, but not long after he decided that mar-riage might improve his situation. It did not. During the summer of 1936 he once again was admitted to a mental facility. His physician thought a trip to Europe might do some good. Carothers heard his mentor and Dupont special consultant, Roger Adams, would be traveling there as a delegate to a major chemical confer-ence. Maybe being around him would assist [24].

---

[13]In the *Roger Adams Papers*, 1937–38 material, there is a Carothers folder in Box 7, Record Series 15/5/23 that discusses Adams knowledge of Carothers challenges.

Fig. 2.4 Wallace Carothers with nylon (Photo courtesy of Getty Images)

Joining Adams and his longtime colleague from his Harvard days, James F. Norris, Carothers was slated to hike with the pair for a couple of weeks.[14] We know that Adams counseled the fellow when they were alone and attempted to encourage him to resume his work. As Adams reasoned it, if Carothers could reflect on all he had accomplished, then it would provide him with the lift he would need to move forward. Somewhere during the hike in the Alps something broke for Carothers, and he made the decision to split off from his companions. Maybe he wanted time to himself in such a scenic spot or maybe Adams' presence was just too much of a reminder of his profession. Adams' demeanor with students was traditionally built

---

[14]James Flack Norris (1871–1940) was an American chemist, who was born in Baltimore, Maryland. Norris was educated in Baltimore and Washington, D.C. before studying at Johns Hopkins University, where he graduated with an A.B. in Chemistry in 1892. He remained at the university to work as a Fellow until 1895, when he was awarded his Ph.D. and became an academic at the Massachusetts Institute of Technology (MIT). He left in 1904 to become the first Professor of Chemistry at the newly formed Simmons College, before returning to take up the position of Professor of Organic Chemistry and, after its creation in 1926, the first Director of MIT's Research Laboratory of Organic Chemistry. Norris served as President of the American Chemical Society from 1925 to 1926 and as Vice-President of the International Union of Pure and Applied Chemistry (IUPAC) from 1925 to 1928.

on heritage and consultancy. So, it was doubtful that he ridiculed or chided his former Ph.D. What was clear was that Adams would never see Carothers again. The chemist with so much promise and ability did not communicate with anyone about when and where he would appear next. He suddenly arrived at his wife's desk at DuPont months after he hiked with Adams. Once again, he checked himself into a mental facility. Not long after that he took his own life with a cyanide capsule that he had carried with him for some time. He knew he could amplify the dose by adding an acid, like lemon juice, and thus he was found dead in a hotel in Philadelphia in the spring of 1937 [28].

Roger Adams rarely spoke of his bond with Wallace Carothers; but when he did, the tenor had an air of reluctance to decipher exactly what happened to him. On the one hand, he really believed that chemists, like that kid from Iowa, developed themselves on their own with only their wits and intellect as a guide. Conversely, he also thought that the power of encouragement would mold someone into a contributing member of the field. Simply being able to look up information in a library did not spell guaranteed success, in Adams' mind. Rather what was necessary was to develop the ability to reason and solve problems by utilizing one's "inborn characteristics," as he liked to say. Carothers was a special person that Adams believed was "unusually good" [29]. And, that was what was so insidious about the man's death. What if there was a drug that could have helped Carothers? Clearly, the medication he was prescribed did nothing to quell his depression and suicidal thoughts. What if Roger Adams, one of the most accomplished organic chemists in the world, could spearhead the research to produce a potentially life-changing pain relief cure? Shortly after Carothers death, he started such a project with the assistance of the federal government. His research focused on unlocking the chemical complexities of none other than that product of hemp—popularly known as marijuana. The rise of CBD was near.

References

1. Leopold A, Leopold L (1993) Round river: from the journals of Aldo Leopold. Oxford University Press, New York, p 146
2. Adams R (1952) William Albert Noyes, 1857–1941. Biogr Mem (Natl Acad Sci U S A) 27:177–208
3. Nutton V (2013) Ancient medicine, 2nd edn. Routledge, New York
4. Boyle E (2013) Quack medicine: a history of combating health fraud in twentieth-century America. Praeger, Santa Barbara
5. Estes JW (1986) Public pharmacology: modes of action of nineteenth-century 'patent' medicines. Medical Heritage 2:218–228
6. McTavish J (2004) Pain and profits: the history of the headache and its remedies in America. Rutgers University, New Brunswick
7. Ludvigsen J (2002) Malt og hamp til børn og svagelige [Malt and cannabis for kids and weakly persons]. Bryggeriet Her
8. Thygesen E (1967) Er marihuana skadelig? [Is marihuana dangerous?] Stig Vendelkjsrs Forlag, Copenhagen

9. Stearns PN (2013) The industrial revolution in world history, 4th edn. Westview Press, Boulder
10. Mann C, Plummer M (1991) The aspirin wars: money, medicine, and 100 years of rampant competition. Knopf, New York
11. Gabriel J (2014) Medical monopoly: intellectual property rights and the origins of the modern pharmaceutical industry. University of Chicago Press, Chicago
12. Liebenau J (1987) Medical science and medical industry: the formation of the American pharmaceutical industry. Johns Hopkins University Press, Baltimore
13. Chandler AD (2005) Shaping the industrial century: the remarkable story of the evolution of the modern chemical and pharmaceutical industries. Harvard University Press, Cambridge
14. Liebenau J (1990) Paul Ehrlich as a commercial scientist and research administrator. Med Hist 34:65–78
15. Rooney S, Campbell JN (2017) How aspirin entered our medicine cabinet. Springer Briefs in the History of Chemistry. Springer, Heidelberg
16. Steen K (2014) The American synthetic organic chemicals industry: war and politics, 1910–1930. University North Carolina Press, Chapel Hill
17. Noble D (1977) America by design: science, technology, and the rise of corporate capitalism. Alfred A. Knopf, New York
18. Kohler R (1991) Partners in science: foundations and natural scientists, 1900–1945. University Chicago Press, Chicago
19. Kohler R (1982) From medical chemistry to biochemistry: the making of a biomedical discipline. Cambridge University Press, Cambridge
20. Adams' Speeches, What the chemist has done for the physician, Roger Adams Papers, University of Illinois Archives
21. Steen K (1995) Confiscated commerce: American importers of German synthetic organic chemicals, 1914–1929. Hist Tech 12:261–285
22. Lenoir T (1997) Instituting science: the cultural production of scientific disciplines. Stanford University Press, Stanford
23. Adams' Publications, The chemistry of synthetic drugs, Roger Adams Papers, University of Illinois Archives
24. Tarbell DS, Tarbell T (1981) Roger Adams: scientist and statesman. American Chemical Society, Washington D.C.
25. Tarbell DS, Tarbell T (1982) Roger Adams, 1889–1971. Biogr Mem (Natl Acad Sci U S A) 53:1–47
26. Adams' Publications, The 50 years of chemistry in the U.S., 1914–1964, Roger Adams Papers, University of Illinois Archives
27. Doel R (2004) Roger Adams: linking university science with policy on the world stage. In: Hoddeson L (ed) No boundaries. University Illinois Press, Champaign-Urbana
28. Hermes M (1996) Enough for one lifetime, Wallace Carothers, Inventor of Nylon. Chemical Heritage Foundation, Danvers
29. Adams' Manuscripts, correspondence with Wallace Carothers' friends, associates, and family, Roger Adams Papers, University of Illinois Archives

Chapter 3
Part Two: CBD and the Tying of Later Bonds

3.1 The Stillborn Birth of CBD

The future may bring a series of drugs that will permit deliberate molding of a person, mentally and physically. When this day arrives the problems of control of such chemicals will be the concern of all. They would present dire potentialities in the hands of an unscrupulous dictator.

—Roger Adams, newspaper editorial, c. 1950s [1].

3.1.1 Government Chemistry Incarnate

In 1938, Joseph Ames was poised to become one of the most important scientists in the America. Then he became ill. In the process, he had to stepdown as chairman of what was at the time a new scientific war-making mechanism for the United States. In that moment, the person who assumed his position was an ambitious and forward-thinking engineer from the Massachusetts Institute of Technology and the private sector named, Vannevar Bush (Fig. 3.1).[1] It is not too strong to argue that his influence on government science would change the world. The organization he now led, the National Advisory Committee for Aeronautics (NACA), was the forerunner to NASA, but at the time it would attempt to harness the most important technological possibility—airpower. Bush understood that science had the greatest potential for human change. In the past, America had a poor track record of getting

---

[1] Vannevar Bush (1890–1974) was an American engineer, inventor, and a science administrator during the Second World War who headed the U.S. Office of Scientific Research and Development (OSRD), which included the early administration of the Manhattan Project. He was chiefly responsible for the movement that led to the creation of the National Science Foundation, and work in early computers for Raytheon.

© The Author(s), under exclusive license to Springer Nature Switzerland AG 2020
J. N. Campbell, *Bonds That Tie: Chemical Heritage and the Rise of Cannabis Research*, History of Chemistry, https://doi.org/10.1007/978-3-030-60023-5_3

civilian thinking to merge with government funding. Afterall, during the First World War, America barely arrived on the Western Front with any supplies. Herbert Hoover's Food Administration stateside was a lone bright spot. Most of the American fight came from hand-me-downs or did not exist at all. The engineer inside Bush wanted to fix that, especially since the size of New Deal expansion had grown federal assistance tenfold. His solution was not to stop with the NACA, but rather to think bigger—he needed a lever [2].[2]

That window opened once Nazi Germany invaded France in May of 1940. The very next month he parlayed his new chairmanship into wrangling a meeting at the White House with Franklin Roosevelt. Gaining access by using his connection to the President's uncle, Frederic Delano, he got FDR to sign off on the National Defense Research Committee (NDRC) in only 15 min. Thus, a piece of paper, allowed Bush to then appoint four leading scientists and old friends to steer the entity: Karl Taylor Compton (President of MIT), James B. Conant (President of Harvard University), Frank B. Jewett (President of the National Academy of Sciences and chairman of the Board of Directors of Bell Laboratories), and Richard C. Tolman (Dean of the Graduate School at California Institute of Technology); Rear Admiral Harold G. Bowen, Sr. and Brigadier General George V. Strong to represent the military. It was a revelatory moment for American science because this organization would in turn propel Bush to lead an even larger and more powerful decision-making government branch—the Office of Scientific Research and Development (OSRD). This organization would direct some of the most significant scientific research of any century; namely the Manhattan Project [3].

The list of bureaucratic "begats" illustrated a massive shift in government-directed science in the United States. During the 1930s before the Second World War began, Bush began to craft what he believed would be the future of this massive field. The "linear theory" he fashioned would shape the Post-War era after 1945. It held that basic research would stimulate and proliferate applied developments. High-caliber programs would be necessary, but there would be little room for improvisation. What Bush was promoting was essentially a new type of scientist that could be sponsored by a government serving the greater good [4]. Previously, as he deemed it, research meandered and lacked a cohesive function. The antidote was the idea that government incentives provided a scientific compass. For instance, you could as a scientist try to solve problems in front of you, like boll weevils devastating crops in the Midwestern America or solve a disease like

---

[2]The New Deal was a series of programs, public work projects, financial reforms, and regulations enacted by President Franklin D. Roosevelt in the United States between 1933 and 1939. The purpose was to provide relief, reform, and recovery from the Great Depression. Major federal programs and agencies included the Civilian Conservation Corps (CCC), the Civil Works Administration (CWA), the Farm Security Administration (FSA), the National Industrial Recovery Act of 1933 (NIRA) and the Social Security Administration (SSA). Support for farmers, the unemployed, youth and the elderly were centerpieces. The New Deal included new constraints and safeguards on the banking industry and efforts to re-inflate the economy after prices had fallen sharply. New Deal programs included both laws passed by Congress as well as presidential executive orders during the first term of the presidency of Franklin D. Roosevelt.

Fig. 3.1 Vannevar Bush, c. 1935 (Photo courtesy of Getty Images)

COVID-19 after it is already rampant, but longer-term solutions were also just as significant to society at-large.

One of Bush's closest allies in the development of this new science was James Conant. And, in turn, one of Conant's closest allies was Roger Adams. Bush had followed Adams work from afar and knew that chemistry, and all its growing power within the scientific community, had great potential. Chemistry, Bush thought, was intimately connected to medicine, war, and thus, life. In order to build a government-focused scientific community, state-led science would need a strong cadre of members with university backgrounds in the lead. There would be many battlefronts to wage war against, particularly in the field of medicine. To win massive global conflicts, it would take research programs that were sanctioned by the federal government and geared toward a specific goal—i.e. fighting pain or curing disease. By the late 1930s, Bush and his team of cohorts were essentially in charge of a developing science. Standing guard alongside them was an equally devoted watchdog that was committed to upholding what they believed was the moral framework of America. They had already taken on organized crime a decade before, now they became even more resolute in policing drug-trafficking. "They" were the Federal Bureau of Investigation (FBI) and the Federal Bureau of Narcotics (FBN), and among their many suspicions, two seemingly unrelated topics piqued their interest [5]. First, was enforcing the newly passed congressional action against the use of marijuana in America. And the second, was the chemical research of a University of Illinois professor named Roger Adams.

3.1.2   "Marihuana:" How Hemp Became Unhealthy

Fifty-eight-year-old Samuel Caldwell was one of the first people to be prosecuted. He was arrested for selling marijuana on October 2, 1937, just one day after passage of the Marihuana Tax Act. Caldwell, a career criminal, was sentenced to four years of hard labor in the most ironic location in America—Colorado. Back then, the state was progressive, just not when it came to legalization of the plant that would become the most controversial in all the land.[3] After 1900, within a matter of decades, marijuana went from medicinal panacea to somewhere between "reefer madness" to the drug that was viewed by some as the concoction of a subversive group of ethnic minorities. How was this possible that cannabis, popularly known as marijuana, could become so unhealthy? The short answer is not per se a chemical explanation. Rather it stems from something all-too-present in the decision making concerning the use of drugs—fear [6].

During the 1930s, the federal government accelerated its expansion. It was a phenomenon that was sweeping the globe. Societies had crashed economically and politically, and in the wake of what would be a Global Great Depression, monies would be funneled into new agencies that were specifically built for creating jobs and scientifically quantifying data. A new age of solving poverty and problems for people brought socialism to national levels. The New Deal attempted to regulate pretty much everything from farms to industrial businesses to Wall Street. Deemed alphabet soup because of its use of acronyms like AAA (for Agricultural Adjustment Act) or NRA (National Recovery Act), these programs were FDR's strategy for stimulating growth and getting the country back on its feet [7]. Some of it worked, but most of it failed in the short term. In the end, only a massive conflict built on fighting Fascism in Europe and the Pacific would help get the American economic juggernaut moving.

Science during the 1930s was one of the areas that reaped the benefits of expanding government. Data-gathering helped make the case for decision making, and chemistry was central to laboratory sponsorship. Quantified studies in a factory-like manner were scaled by Bush and his squads of committees. Universities sought federal funding in an effort to promote their professors—tax money gave. Government agencies came alive again based on the models of the old Bureau of Chemistry from the Progressive Era when the first Roosevelt was President. One of the central foci for government study was to improve life, and what better means to do that than to examine the use of pharmaceutical

---

[3]Colorado Amendment 64 was a successful popular initiative ballot measure to amend the State Constitution; this summarized a new drug policy for cannabis. The measure passed on November 6, 2012, and with the passage of a similar measure in Washington state, marked "an electoral first not only for America but for the world." As of April 2017, 176 of Colorado's 272 municipalities have opted to prohibit retail marijuana activity within their boundaries. While the state's second most populous city, Colorado Springs, has outlawed the sale of recreational marijuana, the city has permitted medical marijuana dispensaries.

manufacturing. How drugs were made and what impact they had on society at-large became a major focus of the new Administration.

Up until this point, pharmaceuticals developed at a breakneck pace. But, alongside the advancements of everyday drugs like aspirin were habit-forming chemicals like heroin, and the concern before 1930 was scattered to say the least. There was the Harrison Narcotics Act of 1914, which was the first comprehensive federal legislation, but enforcement was a challenge since America was preoccupied with staying out of the First World War.[4] Afterwards, alcohol consumption and Prohibition was much more prolific. Drugs like marijuana were not included in the Harrison Act list from 1914 anyway, and that concerned some, but only as a means by which to explain the behavioral patterns of ethnic minorities. Certainly, there were governmental efforts to understand the effects of marijuana and its uses [8]. It was the American pharmaceutical industry that still considered cannabis an important component for medicinal uses, and that may have contributed to it staying off the list. But, that phenomenon would not last.

Enter Harry Anslinger, who was named the commissioner of the Federal Bureau of Narcotics (FBN) in 1930 under the auspices of the Department of Treasury. His reign as the Nation's first drug czar started out fairly innocuous (Fig. 3.2). A former gumshoe and minor diplomat, he arrived in Washington during the dark days of the Hoover Administration. After FDR's election in 1932, the budgets were slashed and the formerly ripe office that Anslinger occupied seemed threatened. So, he did what any government bureaucrat would do in the face of extinction—he invented a reason to exist. The easiest means to combat losing $200,000 in funding was attacking a drug that had a long set of uses among a cross-section of America. Marijuana thus became public enemy No. 1 for Anslinger and his team. The FBN funded focus groups in 1936 and used the American Press to portray the drug as connected to madness [9]. To us in the twenty-first century, this is a familiar argument of how certain drugs corrupt the minds of young people.[5]

---

[4]The Harrison Act was complemented by other Progressive Era laws (for example, the Uniform State Narcotic Drug Act in 1934). The number of addicts of opium started to decrease from 1925 to a level that in 1945 that was about one tenth of the level in 1914. The use of the term "narcotics" in the title of the Act was used to describe not just opiates but also cocaine—which is a central nervous system stimulant, not a narcotic. Today, the term "narcotic" translates to any illegally used drug, but it refers to a controlled drug in a context where its legal status is more significant than its physiological effects. The remaining effect of the Harrison Act has been supplanted by the Controlled Substances Act of 1970.

[5]Harry Jacob Anslinger (1892–1975) was a United States government official who served as the first commissioner of the U.S. Treasury Department's Federal Bureau of Narcotics during 5 presidencies. He was a supporter of Prohibition and the criminalization of drugs. He victimized ethnic minorities, while spreading anti-drug policy campaigns. Holding office an unprecedented 32 years in his role as commissioner until 1962, Anslinger held office two years as U.S. Representative to the United Nations Narcotics Commission. The responsibilities once held by him are now largely under the jurisdiction of the U.S. Office of National Drug Control Policy.

Fig. 3.2 Narcotics Commissioner Harry J. Anslinger, testifying before the Senate Judiciary Subcommittee in 1955 (Photo courtesy of Bettmann/Getty Images)

Reefer Madness (originally made as *Tell Your Children,* and sometimes titled as, *The Burning Question or The Dope Addict,* just to name a few) is a 1936 American film that reached a broad audience (Fig. 3.3). Hoping to escape the daily rigors of the Depression, it was affordable and almost every major town had a theater. Messages were communicated both directly, and not quite so, in forms utilizing allegory and metaphors. *Reefer Madness* was as direct a story as possible. High school-age students were susceptible to the trappings of an evil substance and consumption would lead to nothing short of death, according to publications that supported the message [10].[6]

Priming the popular culture pump, Anslinger laid out a plan to target marijuana and have it added to the list of federally banned substances. These would be administered under the authority of the FBN. Chemists and medical personnel were still divided over its effects, but the legal justification for banning it was already in place, as two-thirds of state legislatures had already moved to ban it by the early 1930s. As mentioned, racial divisions were resolute; justifying stiff penalties and in place to check crime rates from rising to intolerable levels. Previously, marijuana had flowed through ports like New Orleans as sources in the Caribbean and beyond continued to propagate it. Global organizations, like the powerless League of Nations, also assumed previous drug-related councils to legislate on a wider scale. The International Narcotics Control Board, which still exists today within the

---

[6]Editorial control for the film was heavily influenced by the FBI, directly.

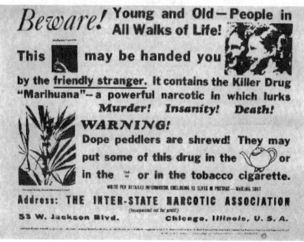

Fig. 3.3 Popular culture and the evils of marjijuana use—the film, *Reefer Madness,* and government advertisement from a unknown magazine, both mid-1930s (Photos courtesy of the author)

United Nations structure, was such an entity [11]. By 1936, they were also zooming in on marijuana, making the arguments that there was substantial evidence that using it would lead to widespread moral depravation.

As western ideas continued to reform and adjust to the idea that cannabis and its products were reprehensible, Anslinger used his rising influence to prod the Congress to act, which they did. In 1937, the Marihuana Tax Act, that prosecuted Samuel Caldwell, was introduced by Representative Robert L. Doughton of North Carolina, ironically a hemp-producing state. Rushed through the bill process, it levied harsh penalties against perpetrators, as Samuel Caldwell found out. Historical arguments make interesting claims concerning motivations for the Act beyond the influence of Anslinger. Some have thought the move stemmed from the influence of the controversial newspaperman William Randolph Hearst, who did not want hemp (hurds, as they were dubbed) to rival wood pulp and devalue his timber investments [8]. Others claim the du Pont Family, the owners of Wallace Carothers' contribution to nylon, were concerned that cheap hemp would replace this new product. None of these hold much water and judging by Anslinger's rhetoric, it was his influence that was paramount.

Not everyone was sold on Anslinger's gambit for the national control of marijuana usage. The powerful Mayor of New York City, Fiorella LaGuardia, smelled a rat.[7] In 1938, he formed his own committee to report on the state of marijuana usage in his city after claims that the drug was readily available in schools. He added chemists, as well as psychological specialists, to a group that would examine every aspect of the dilemma. It was a ground-breaking study that received little press attention at the time. The Mayor stated an interesting hope for the future by proclaiming that, "The scientific part of the research will be continued in the hope that the drug may prove to possess therapeutic value for the control of drug addiction." Published officially towards the end of the Second World War in 1944, below are some of those "scientific" conclusions that were reflective of attitudes of the period. Most affirmed previous notions asserted by members of the medical community and would become defenses for cannabis legalization in the future [6]. Conclusions listed in the summary and discussion of the report included [12]:

1. Under the influence of marihuana the basic personality structure of the individual does not change but some of the more superficial aspects of his behavior show alteration.

---

[7]Fiorello Henry La Guardia (1882–1947) was an American politician, best known for being the 99th Mayor of New York City for three terms from 1934 to 1945. He was a pro-New Deal liberal progressive Republican. Previously he had been elected to Congress in 1916 and 1918, and again from 1922 through 1930. Temperamental and charismatic, he appealed across party lines. As a New Dealer, he supported President Franklin D. Roosevelt, and in turn he heavily funded the city, thus severing patronage for the Mayor's enemies. La Guardia revitalized New York City and restored public faith in City Hall. He unified the transit system, directed the building of low-cost public housing, public playgrounds, and parks, constructed airports, reorganized the police force, defeated the powerful Tammany Hall political machine, and reestablished employment on merit in place of patronage jobs.

2. With the use of marihuana the individual experiences increased feelings of relaxation, disinhibition and self-confidence.
3. The new feeling of self-confidence induced by the drug expresses itself primarily through oral rather than through physical activity. There is some indication of a diminution in physical activity.
4. The disinhibition which results from the use of marihuana releases what is latent in the individual's thoughts and emotions but does not evoke responses which would be totally alien to him in his undrugged state.
5. Marihuana not only releases pleasant reactions but also feelings of anxiety.
6. Individuals with a limited capacity for effective experience and who have difficulty in making social contacts are more likely to resort to marihuana than those more capable of outgoing responses.[8]

Anslinger and LaGuardia represented two halves of a political coin that would divide the nation over the question of, legal or illegal, healthy or unhealthy? For the time being, the former controlled the moral high ground by using political connections as a means to an end. While the latter argued that not everyone was so sure that marijuana, hemp, or cannabis, depending your usage and taxonomy, was harmful. Chemists were especially interested in its properties. In particular, the chemist who supplied LaGuardia's study with actual marijuana capsules and the placebos was a professor of chemistry at the University of Illinois; none other than Roger Adams.[9] Here is what he did.

3.1.3 Adams' Laws of Extraction

By 1938, if you wanted to ingest or inhale marijuana for any purpose, you could obtain it around the globe in a variety of ways; however, in America it was becoming more difficult. Like the gun and gambling laws, which were also being overhauled at both the national and local levels, marijuana was being segregated. You could find it on the street as the drug retreated into the shadows or you could · grow it yourself. There was the US Drug Enforcement Administration (DEA) with aerial capability back then (it was not installed until the Nixon Administration), but as mentioned, there was the Department of the Treasury and its Bureau of Narcotics. Obtaining the plant-based materials was not out of the question, but

---

[8]Even though the LaGuardia Commission's Report was squashed in the popular press at the time, it has since been rediscovered time and time again by cannabis supporters since the 1970s. *Summary and Discussion.* LaGuardia Commission Report. http://druglibrary.net/schaffer/Library/studies/lag/sumdis.htm.

[9]In the 1944 LaGuardia Report, Roger Adams' name is mentioned several times as a consultant. He supplied many of the "control" products. "For controls, glycyrrhiza pills without marihuana were used. Several products prepared by Dr. Roger Adams in his investigation of the chemistry of marihuana were used. A comparison of their action with that of the concentrate will be found below." https://daggacouple.co.za/wp-content/uploads/1944/04/La-Guardia-report-1944.pdf.

American culture was starting to question its previous usage for medicinal purposes, and instead began to think of the drug as both illegal and as a gateway to malfeasance.

During this time period, as governments struggled to maintain a modicum of sanity, Roger Adams was focused on becoming a chemical juggernaut. His laboratory at the University of Illinois was top-tier when pertaining to productivity, and his organic chemistry program focused on the occurrence of hindered rotation in biphenyls and related compounds (Fig. 3.4). Publishing over 60 papers by the early 1940s, some of his critics argued that this research and reporting led by him was becoming recycled over time. Adams disagreed, claiming that it was an excellent opportunity to teach the chemists of tomorrow about how important the fundamentals truly were—so he did not bow to pressure; he was an avenger, so he pushed on [13]. Next, he tackled gossypol in the mid-1930s, which was a highly toxic cousin of cottonseed that had a complex structure never defined by previous laboratories. Adams' Champagne-Urbana Team set to work, using ultraviolet spectroscopy for the first time in these types of investigations. Results yielded a series of masterfully complex papers, and probing, in turn, spread to different sectors of the chemical world after 1945. Currently, gossypol is prescribed to men as a birth control, while women take it for disorders of the uterus including endometriosis, abnormal bleeding, and ovarian cancer. Adams was not done yet [14].

Most historians who have analyzed what rationales Roger Adams applied when he took up the proposition to unlock the secrets of marijuana seemed convinced that he was motivated by two forces. First, his research and extracurricular activities as a consultant during the late 1920s and early 1930s were focused on pharmaceutical research. This penchant for assisting medical personnel in their fight against pain and suffering was something Adams was committed to after he witnessed the power of chemical connections in Germany before the First World War. Thus, some premises have clearly stated that Adams knew marijuana studies were significant, but they emphasize that he was the one that wanted to see research conducted as soon as possible before other American or British labs could scoop him. Second, they seem rather taken with the idea that the Chief was "asked" to conduct studies by the Treasury Department and the Bureau of Narcotics under Harry Anslinger sometime around 1939–1940 [15]. This, they surmise, was a request channeled, so the federal government could use this research to bolster their case to continue the banning of this illegal substance.[10]

There is no doubt that the first part of these assertions is accurate. Adams was focused on what the chemist could do for the physician, but a caveat to consider was that he never felt that being first on some research goldmine made his claim

---

[10]Two Adams scholars that put forth this argument that the chemist's motivations for studying marijuana were more professionally based include: D. Stanley Tarbell and Ann Tracy Tarbell, Roger Adams: Scientist and Statesman (1981), and Ronald E. Doel, "Roger Adams: Linking University Science with Policy on the World Stage," in No Boundaries: University of Illinois Vignettes, Lillian Hoddeson, ed. (2004). The Tarbell's monograph is the only biography of Roger Adams ever completed.

Fig. 3.4 Roger Adams, post-1945 (Photo courtesy of Edgar Fahs Smith Collection, Kislak Center, University of Pennsylvania)

more important than any others. As mentioned, he was the consummate mentor/ collaborator, and after the sudden suicide of his brilliant student, Wallace Carothers, he was never the same. It troubled him that Carothers clearly could not receive the help he needed. His relentless pursuit of meaningful projects led him to examine marijuana as a possibility for pain relief. He was particularly interested in its effects on the brain, not competition from rival labs [16].

Once Adams heard that the Marihuana Tax Act was a reality, he followed the story, all the while knowing that there were limits to what he could accomplish. Marijuana already had a complex history, but maybe if it could be proven through scientific inquiry that cannabis sativa possessed important components, then a case could be made for its own validity. Historical motivations are always more complicated. That is the main reason why Adams requested that the Treasury Department send him samples of the stuff.[11] His intent was to test it for a new study which was set to probe the chemical, psychological, legal, and societal aspects of the drug—the LaGuardia Report. If historians had examined this document more

---

[11]Some chemical articles, with a historical bent, point to this phenomenon of laboratories requesting to test illicit drugs as being something that dates from the 1970s. Adams' example reflects an earlier interest. Clearly, chemists were requesting samples a generation before.

closely, they would have noticed that Adams' name was prominently referenced as part of the survey. In other words, what his team extracted in the end would be of interest to Harry Anslinger, but he was not going to be very pleased with Adams [17].

3.1.4   Enter Cannabidiol 1.0

As the complex history of cannabis reflects, unsuccessful attempts to discover what was known as an "extract" abounded in the nineteenth and early twentieth centuries. These were all the rage as laboratory expansion professionalized and prospered after centuries of these types of liquids emanating from cooks' kitchens. Old medicinal lines were crossed, as we saw with Maltos Cannabis, and new products were created. By the late 1930s, pharmaceutical sales became a major component of the economy—the sky was the limit. Roger Adams closely followed developments and knew his history. He understood that Wood, Spivey, and Easterfield had extracted hemp with alcohol and then distilled the solution under reduced pressure to obtain a high boiling fraction as a red oil (Fig. 3.5).[12] This red oil was then further treated with acetic anhydride to give two products—a crystalline acetyl derivative and a residual oil, which was also thought to be an acetyl derivative. The former was determined to be that of cannabinol (CBN). None of this was CBN, however, but derivatives of it [18].[13]

Thus, speaking of stillborn births, cannabinol was a "physiologically active substance," so it was not thought to be the active principle in marijuana. Later investigations, particularly in the early 1930s before Adams was involved, yielded new results with the research of British chemist Robert S. Cahn.[14] Before that point,

---

[12]Wood, Spivey, and Easterfield were a group of Cambridge University chemists who in the 1890s succeeded in obtaining a relatively pure extraction of cannabis which they called "cannabinol" (See, Wood TB, Spivey WTN, Easterfield TH (1899) Cannabinol: Part 1, J. Chem. Soc. Trans, 75:20–36). In a related experiment, sadly Spivey and Easterfield were killed in a laboratory explosion. Wood almost suffered the same fate when he swallowed cannabinol and lost consciousness, a fire ensued. If it were not for a nearby lab assistant who smelled smoke and rescued him, he would have certainly died. See, Ernest L. Abel, *Marihuana: The First Twelve Thousand Years* (1980), 170.

[13]The acetyl derivative could be saponified (i.e. removing the acetyl group) and the residue extracted with ether. Distillation of this resulted in an "almost colourless oil" that was determined to be cannabinol.

[14]Robert Sidney Cahn (1899–1981) was a British chemist, best known for his contributions to chemical nomenclature and stereochemistry, particularly by the Cahn–Ingold–Prelog priority rules, which he proposed in 1956 with Christopher Kelk Ingold and Vladimir Prelog. Cahn was born in Hampstead, London. His best-known work concerning cannabidiol can be found, Cahn RS. 1932. Cannabis indica resin. III. Constitution of cannabinol. J Chem Soc 1342–53.

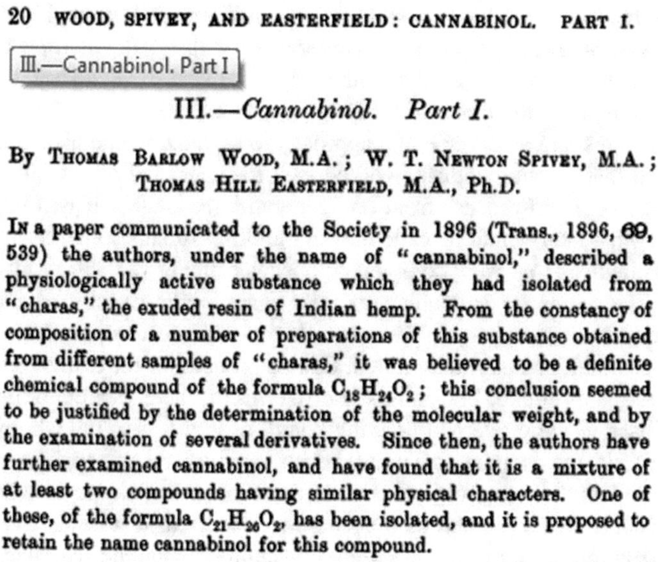

20 WOOD, SPIVEY, AND EASTERFIELD: CANNABINOL. PART I.

III.—Cannabinol. Part I

III.—*Cannabinol. Part I.*

By THOMAS BARLOW WOOD, M.A.; W. T. NEWTON SPIVEY, M.A.;
THOMAS HILL EASTERFIELD, M.A., Ph.D.

IN a paper communicated to the Society in 1896 (Trans., 1896, 69, 539) the authors, under the name of "cannabinol," described a physiologically active substance which they had isolated from "charas," the exuded resin of Indian hemp. From the constancy of composition of a number of preparations of this substance obtained from different samples of "charas," it was believed to be a definite chemical compound of the formula $C_{18}H_{24}O_2$; this conclusion seemed to be justified by the determination of the molecular weight, and by the examination of several derivatives. Since then, the authors have further examined cannabinol, and have found that it is a mixture of at least two compounds having similar physical characters. One of these, of the formula $C_{21}H_{26}O_2$, has been isolated, and it is proposed to retain the name cannabinol for this compound.

Fig. 3.5 Early report of cannabinol, late nineteenth century (Photo courtesy of the author)

red oil was in hyper-sleep, with few assessments that yielded results.[15] Maybe it was the series of conflicts afoot or the advent of other drugs from aspirin to antibiotics, we are not sure. Whatever the case, Cahn advanced the research enough to declare that a position existed close to hydroxyl and the *n*-amyl groups. Yet, something was missing—the orientation of all the groups [13]. That was Adams' window.

Once the LaGuardia Study went live, Adams' laboratory in turn contacted the Treasury Department. This raised a few eyebrows, but Anslinger was willing to go along since he was confident that no one could prove that marijuana or any derivates would be useful for treatment. Adams had other ideas, as he said in a major paper that he would later present at the Harvey Lectures, "The primary objectives were the determination of the mental and physical actions of marihuana on the kind of person resorting to its use and the consequent social implications." Placing the full power of his laboratory structure behind his efforts to properly unlock the

---

[15]Several chemical papers and investigations were conducted in the early twentieth century, but none could be described as breakthroughs. These include: Czerkis M (1907) Beiträge zur Kenntniss des cannabinols, des wirksamen bestandtheils des haschisch. Justus Liebigs Annalen der Chemie 351:467–72; Bergel F (1930) Einige Beiträge zur konstitution des cannabinols, des wirksamen prinzips im haschisch. I. Justus Liebigs Annalen der Chemie 482:55–74, to name just a few examples.

secrets of marijuana, Adams postulated that hydroxl and *n*-amyl groups must occupy positions different from those suggested by Cahn [19]. Attempts would turn into success very quickly as the University of Illinois' chemistry laboratories suddenly came upon the dawn of CBD.

Roger Adams had a very specific approach to organizing chemical laboratories. He believed that certain chemists worked well together, and others did not. As he imparted heritage and process from one generation to another it was important to emphasize what later would be called, team building. Adams sometimes found that when two chemists that were complete opposites came in contact, results could be obtained. The teams he assembled to investigate where exactly cannabinol fell short, started with samples of Minnesota hemp. The first experiments were not terribly successful; actually, they failed. Cahn's procedures for extraction could not be replicated, they found; but, in this abortive result, there was opportunity.

Going back to the proverbial drawing board, or in this case the lab, by utilizing qualitative tests, the chemists, 10 in all, began to examine opportunities to isolate a phenolic product.[16] The result was one of those chemical moments where break-throughs occur. Of the numerous reagents that occurred, it was discovered that a dinitrobenzoyl chloride reacted to give a crystalline compound the chance to then be removed and purified. The result proved to be what Adams and the chemical world called a bis-ester. Through hydrolysis, a new substance formed from the presence of two phenolic groups that the Adams' squad called, cannabidiol, more commonly referred to now as CBD [20].

At the time, what was so revolutionary about both cannabidiol and cannabinol is that were the only pure compounds related by structure to the active constituents which were made from hemp extracts. Chemical derivatives, through the processing of oils became crystallized more fully with future research by the team. Probably what was most far-reaching was the summation of cannabidiol's rouse. Although it is physically inactive, like its cousin cannabinol, there were some revealing dif-ferences. First, Adams's chemical coterie presented convincing evidence that cannabidiol was composed of dihydrocymyl and olivetol (I, 3-dihydroxy-5-n-amylbenzene) residues. After reduction to tetrahydrocannabidiol, and then oxi-dized, menthane carboxylic acid was then isolated. Adams' lab did not have enough time to explore the stronger of the two. They believed that it was likely that cannabidiol was a precursor of cannabinol. In the end, plant reactions became the key. Tetrahydrocannabinol formed by isomerization of cannabidiol with various acidic reagents being obtained in two forms depending on the reagent and physical

---

[16]Adams committed himself to the promotion of his students. He always strove to give them credit and assist in seeking outlets for their publishing as soon as possible. B.R. Baker, C.K. Cain, J.H. Clark, Madison Hunt, Charles Jelinek, W.D. McPhee, D.C. Pease, C.M. Smith, R.B. Wearn, and Hans Wolff, all received publishable credit and were part of the final process when Adams released his Harvey Lectures in 1941–1942.

conditions used in the reaction. All this added up to a more complete understanding the components of red oil and its possible uses, especially as a possible pharmaceutical for the future.[17] Adams' lab had a caveat though as their research was ending—it was an open door they were unable to walk through. They made it known that reliable scientific inquiry into the pure active compounds would have to await further study and investigations. More had to be done, but they understood that chemistry was a series of steps and could not be taken all at once. Some of the early clinical tests run by the LaGuardia Report showed promise, but these were on inmates at Rikers Island. Not exactly the kind of clientele that a respected study would include. Politics clouded the chemistry, and it would be that way in America with marijuana and CBD until the twenty-first century.

Adams mused that the properties of these findings could have long-reaching effects on the future of marijuana, but he knew he had to be cautious about his comments. Harry Anslinger and the Treasury Department were listening, and it just so happened that some ill-timed points at a public speech by him drew the ire of the Drug Czar. Actually, once Bush became aware of Adams interest in obtaining security clearance when the Second World War began, he had to intervene on behalf of the "Chief" after the FBI flagged his application. This was not for his supposed "communist" rhetoric and support of left-leaning organizations, but rather due to his support for marijuana research [15].[18] Global conflicts have a way of waylaying love and chemistry, but it seems of the two, science can be undeterred. Bush got Adams involved in his war using science, but CBD experienced a still-born birth. It would have to wait on the power of chemical heritage to be passed down to the next generation. Adams' era was passing. This time though it was the bonds of international chemistry that assisted, albeit a cold one, but not before the branch became the premier wing of science in the post-War era.

---

[17]Adams made several public speeches and published articles in the late 1930s and early 40s about the prospects for marijuana and its possible applications. For a detailed examination of these speeches and published works, see, the *Roger Adams Papers*, 1812–1971, housed at the University of Illinois Archives, https://archon.library.illinois.edu/?p=collections/controlcard&id=3741.

[18]Historian Ronald Doel makes the argument that part of the reason Adams' security clearance was waylaid was his membership in the "Lincoln's Birthday Committee for the Advancement of Science," which was a supposed communist front group from the 1920s. While this was true, the FBI was investigating Adams on this connection, the stronger objection were his research ties to marijuana [15]. And this is what Anslinger's ire, and thus, the FBI's interest in withholding security clearance.

3.2   Chemical Heirs and Marijuana Research

"We will bury you!" (Russian: Мы вас похороним! (Romanized): "My vas pokhoronim!")

—Soviet First Secretary, Nikita Khrushchev, while addressing Western ambassadors at a reception at the Polish embassy in Moscow on November 18, 1956 [21].

To assure the integrity and vitality of our free society, which is founded upon the dignity and worth of the individual. The Soviet Union represents slavery under the grim oligarchy of the Kremlin.

—from NSC-68, the planning document adopted by the Truman administration in 1950 and one of the foundational texts of the early Cold War [22].

3.2.1   Cold War Chemistry and the New Medicine

Alexander Solzhenitsyn, the now celebrated Russian author who endured the Gulag and hospitalization under Russian medical care, had a solid understanding of the mindset of a physician and the power that pain could command over an individual. *Cancer Ward* (Russian: Ра́ковый ко́рпус, Rákovy kórpus) is a semi-autobiographical novel by the Nobel Prize-winning author.[19] Completed in 1966, it was distributed in Russia that year in Samizdat, the secret method for printing objectionable works, and banned there the following year. This Solzhenitsyn-tale relates the story of a small group of patients in Ward 13, a cancer wing of a hospital in Soviet Central Asia in 1955, two years after Joseph Stalin's death. A range of characters are depicted, including those who benefited from Stalinism, resisted or acquiesced [23].

Like Solzhenitsyn, the main character Oleg Kostoglotov, spent time in a labor camp as a "counter-revolutionary" before being exiled to Central Asia under Article 58. In the book, one of the central foci is the "ward round," whose purpose is to improve the morale of the patients. Euphemisms, vague formulations, and down-right lies are routine tools of the trade through the process. The medical personnel

---

[19]Aleksandr Solzhenitsyn (1918–2008) was a Russian novelist, philosopher, historian, short story writer and political prisoner. Solzhenitsyn was an outspoken critic of communism and helped to raise global awareness of the Gulag. After serving in the Soviet Army, he spent eight years in a labor camp, and then internal exile for criticizing Stalin in a private letter. He published only one work in the Soviet Union, the novel *One Day in the Life of Ivan Denisovich* (1962). Although the reforms brought by Nikita Khrushchev freed him from exile in 1956, the publication of *Cancer Ward* (1968), *August 1914* (1971), and *The Gulag Archipelago* (1973) caused the revocation of his Soviet citizenship in 1974. He was flown to West Germany, and in 1976 he moved with his family to the United States. When Soviet Union dissolved, his citizenship was restored in 1990, and four years later he returned to Russia, where he remained until his death in 2008. He was awarded the 1970 Nobel Prize in Literature. *The Gulag Archipelago* sold tens of millions of copies.

rarely say what they think—until they sit down together later and "the general impression of improvement and recovery was completely exploded." One patient who has not responded to treatment is simply ignored by the physicians. Patients are discharged before they can die, to improve the clinical statistics—no palliative care here. During individual consultations some of the physicians are more honest, and even show natural kindness, rather than mere professional courtesy. When Ludmila Dontsova, the lead oncologist, becomes ill with cancer, she does not want to know anything about the details of her condition, treatment, or prognosis—a reflection of those that do not want to look at the Stalinist past, nor contemplate the future [24].

Allegorically, *Cancer Ward* has much to tell us about the state of medicine in Russia after the Second World War, but if we broaden our perspective, it goes much further in explaining what both superpowers (US vs. USSR) missed. That being, as laboratory medicine moved into a new era of development and the world became more interconnected, the poor relationship between both sides stymied the passing along of knowledge and heritage. More specifically, if we examine the role of marijuana research in the wake of the Adams lab's discovery of CBD, both sides limited their explorations because they feared what the other represented politically [25]. The Cold War divided much more than just territories, instead it both empowered state-sponsored science and was a detractor to it.

There is a consensus among historians of science that of all the disciplines, the one that came out of 1945 with the most to gain was chemistry. After all, physics created something in nuclear power that was, as of yet, without practical application. Yes, power could be harnessed using clean energy, but there were all sorts of question marks when it came to safety. Disasters from Three-Mile Island to Chernobyl would plague this resource. Chemistry on the other hand was different. Out of the greatest global conflict in world history, probably what did more to save lives than anything were antibiotics. An army and a people that could fight bacteria, could keep their strengths up, and a society that wielded this power was unstoppable. Under the Marshall Plan, the West did everything possible to hoard antibiotics and keep the stuff out of the hands of the Communist regimes in Europe and Asia. At first, this seemed like the logical course of action, but what the United States did not realize is that they were creating a massive problem over time.

Biochemical experts who a few generations before were shunned by mainstream chemistry were now positioned to advise nations how to invest in agriculture and to promote health. The answer was using antibiotics to treat all sorts of bacteria. We know today that not all of it is "bad," there is much to admire in bacteria and how it assists. Cannabis was one of those parts of agricultural development that was shunned by the United States. Since it was vilified back in the 1930s, its growth was limited. Instead, other plants and medicines were being manipulated and grown using antibiotics instead of vitamins and fertilizers. By the end of the century, new breeds of bacteria began to prosper as a natural occurrence. Chemists tried to keep up with these changes, but there are limits to progress. Over time, pharmaceutical companies began to shift their focus away from antibiotic research, and instead turn their attention to biotechnology. Someday soon, the field of biochemistry that focuses on antibiotics will cease to exist. Agriculture, which for so long saw massive shifts in the production of food, will not be able to keep up [26].

What was it all for, the Cold War? East v. West competition for hearts and minds were battled in a series of proxy wars in seemingly faraway places. Korea, Vietnam, Africa, Central America, and the Middle East were all affected by influences from both the Americans and the Soviets. But the Cold War of science was even more destructive because it was subtle and slow-acting. Gradually, as the world entered the 1960s, chemists began to question whether the new medicine based in antibiotics was the best course of action. Some resisted, much like the bacteria they knew so little about. This new offspring of the old chemistry before the Second World War was different in that it existed outside the lines drawn by the East-West. Freedom gave these scientists the opportunity to open their own investigations, while still drawing on a chemical heritage. Non-aligned, their greatest example was a person who possessed two distinct lineages. His background was very different from Roger Adams. Instead the world he lived in was one that presented a fresh scenario of peril. Yet, he also opened the possibility to transcend politics, in order to move cannabis research into an incredible era of expansion.

3.2.2   Heir 1: The Non-aligned Chemist, Raphael Mechoulam

By the 1960s, with the Cold War in full swing, the history of chemistry reached a new epoch of complexity and hegemony. Global laboratories had grown in strength, especially outside the East-West divide, and this equated to a major boom in productivity. Organic chemists were as active as ever, and they were part of an increasingly powerful network of ideas and options. The discipline now had a major biological component and the organic market bolstered itself with pharmaceutical research as a means for companies to expand across nations to build a diversified workforce to handle demand. Governments that waged wars and conflicts continued to use chemistry for their own ends, and this had unintended consequences. For instance, the Third Reich drove all sorts of peoples from their ancestral homes and this created a migratory pattern that had a major impact on science. After the dust and blood cleared and this new Cold War divided, a new generation of chemists were the first to be trained in rebuilt labs and with the techniques that war wrought [27].

Raphael Mechoulam was one of those chemists who was affected deeply by the war. He was born in Sofia, Bulgaria in 1930, to a Sephardic Jewish family. His father was a physician and head of a local hospital, while his mother hailed from a wealthy Jewish family that could afford to send her to study in Berlin. Mechoulam attended an English-speaking school until Anti-Semitic laws became overtly punitive. When his father was deported to a concentration camp, the family had no idea if he would survive. By some miracle he did, and after the Communist takeover of Bulgaria, the son became interested in science. Interestingly, his first foray was into a related field—chemical engineering, but he did not take to it.[20] When his family immigrated in 1949 to Israel, he began to study organic chemistry and was sold [28]. He gained his first research experience, like many young compatriots, in the Israeli Army working on the chemistry of insecticides, which were undergoing a major revolution in treating a range of pests.

The rest of his ascendancy to scientific stardom started when he received his M. Sc. in biochemistry from the Hebrew University of Jerusalem (1952), and his Ph.D. at the Weizmann Institute, Reḥovot (1958). His thesis covered the chemistry of steroids, which was becoming a major international issue in the wake of Nazi research and its use pertaining to athletic prowess. After postdoctoral studies at the Rockefeller Institute, New York (1959–60), he joined the Weizmann Institute (1960–65), before moving to the Hebrew University of Jerusalem, where he became professor (1972) and was then elevated to the Lionel Jacobson Professor of Medicinal Chemistry in 1975. Later he served as rector (1979–82) and pro-rector (1983–85). The capstone of his career came when he was elected as a member of the Israeli Academy of Sciences (Fig. 3.6) [29].

What Mechoulam did, and is still doing over a long career, was scoop up the baton of chemical heritage. His work on CBD took a major step forward in our understanding of the potential for marijuana research. As an Israeli national, he could not have conducted the extensive chemistry that was necessary during the 1960s, without the liberty that was given in his country. The United States was, and mainly still is, closed to more advanced research. Mechoulam recognized the opportunity before him and picked up where Adams and his team left off. As a non-aligned chemist, that being one who does not belong to either the American or Soviet side, he had the opportunity to experiment outside the confines of Cold War science. The result would have a lasting impact on the history of cannabis research.

Mechoulam started his investigations by wanting to find some means to utilize the ever-expanding methods of Nuclear Magnetic Resonance (NMR) spectroscopy that was becoming more prevalent after 1945.[21] He had heard of some of the tests on marijuana and was more than curious about its chemical properties, and he thought it possessed great potential when it came to possible pain relief and low amounts of side effects. In other words, it had a bad name. Coupled with this zeal,

---

[20]Raphael Mechoulam (1930–) is an Israeli organic chemist and professor of Medicinal Chemistry at the Hebrew University of Jerusalem in Israel. Best known for his work (together with Y. Gaoni, C. Trips, and S. Benezra) in the isolation, structure elucidation, and total synthesis of Δ9-tetrahydrocannabinol; he also examined the main active principle of cannabis and for the isolation and the identification of the endogenous cannabinoids anandamide from the brain and 2-arachidonoyl glycerol (2-AG). He has become an ambassador for the chemical research of cannabis.

[21]The discovery of Nuclear Magnetic Resonance (NMR) spectroscopy is ascribed to Isidor Isaac Rabi, who received the Nobel Prize in Physics in 1944. The Purcell group at Harvard University and the Bloch group at Stanford University independently developed this type of spectroscopy in the late 1940s and early 1950s. While Edward Mills Purcell and Felix Bloch shared the 1952 Nobel Prize in Physics for their discoveries.

Fig. 3.6 Raphael Mechoulam, during a lecture, c. 1964 (Courtesy of Zach Klein, from his documentary, *The Scientist*)

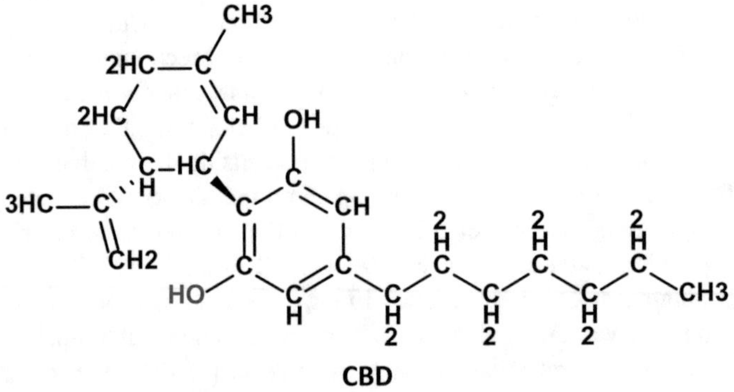

Fig. 3.7 CBD (Courtesy of Steven Rooney)

Mechoulam's work on insecticides and unlocking their secrets was an enticing puzzle—as it would be with cannabis [30].

Like Adams before him, the young chemist from Bulgaria who became an Israeli citizen, hadn't the foggiest idea where to get some. Cannabis was just as regulated in the 1960s as it was in many Western countries. So, Mechoulam did the only thing he knew he could, he reached out to a colleague/mentor who knew members of the police department. Those folks would have access to marijuana that was seized in raids, and as long as Mechoulam obtained the proper permit, he could conduct his research. Ironically, he picked up the first "stash" from the local precinct in a satchel and rode the bus back to university. That is how a new era in CBD began (Fig. 3.7).

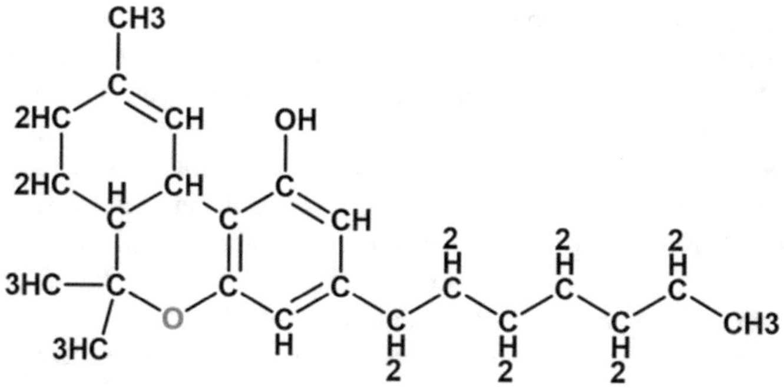

Fig. 3.8 THC (Courtesy of Steven Rooney)

Mechoulam began work, along with two other colleagues, and isolated several of marijuana's 142 active components which led to the mapping of their structures. The goal was to make all of them available for pharmacological and clinical research. This was important. Times in Israel were changing when it came to research and development. Unlike the United States, that could take place and the laws were much looser in this regard. Mechoulam had categorically broken through the wall of cannabis, and with prior data in hand, he unlocked the secrets of these compounds that he would call—cannabinoids. By 1964, his mapping and isolation led to a more thorough understanding of what would become in the nomenclature of everyday life as CBD (cannabidiol) and tetrahydrocannabinol (THC). The latter, it was discovered, switches on the receptors in the body and may relieve pain and even heal Fig. 3.8) [31].

The key to the discovery was uncovering how the body's natural cannabinoid receptors could utilize biologically similar compounds that are already contained within us. More would need to be done, but the path led this time by a non-aligned chemist like Mechoulam, was once again open. This research could not have been led by an East or a Western chemist. Like the characters in *Cancer Ward*, they were too constrained by preconceived notions and the science was in turn hampered by a lack of vision aimed at a common goal. It would take a new breed of scientists to take up the mantle of CBD research, though for Mechoulam, he was not done yet (see Appendix, for timeline).

3.2.3 Heir 2: The Agonist Chemist, Allyn Howlett

While Raphael Mechoulam and his coterie were setting the international chemical cannabis world on fire with their ground-breaking research, American chemists faced several obstacles. The 1970s brought a new level of banishment for cannabis after the counterculture of the 1960s helped to give the plant an even worse name than in prior decades. The Comprehensive Drug Abuse Prevention and Control Act kicked off the new Nixonian decade by brutally ravaging the world's most powerful country's ability to investigate the chemical properties and secrets of cannabis. The aftershocks would resonate to this day and beyond. Controlled substances were divided into five schedules (or classes) these included: the basis of their potential for abuse, accepted medical use, and accepted safety under medical supervision.

To this day, proceedings are naturally overtly elaborate and to add, delete, or change the schedule of a drug or other substance would have to be initiated by the newly created Drug Enforcement Administration (DEA, c. 1973), and the Department of Health and Human Services (HHS). There is another avenue, but it is fraught with bureaucratic pitfalls. Those interested can, by petition, "sponsor" a recommendation. This can come from any interested party, including the manufacturer of a drug, a medical society or association, a pharmacy association, a public interest group concerned with drug abuse, a state or local government agency, or an individual citizen. Once a petition is received by the DEA, the agency begins its own investigation of the drug. As chemists found out, this could take years. Cannabis has been a Schedule 1 from the beginning of the Act initiation [32].

In the wake of this massive government oversight, by the mid-1980s cannabis-related research underwent a series of alterations as chemical data was amassed by new testing initiatives. One of the main focal points was a chemist named Allyn Howlett, who was part of a major influx of women into scientific fields that steadily gained steam after 1960. Trained at Rutgers University, she attained a Ph.D. in 1976, and went on to serve in several key academic and private sector positions. Currently, she serves as a biochemical neuropharmacologist at Wake Forest University's School of Medicine. Howlett, whose expertise includes cannabinoid receptor signal transduction, continues to serve as an agonist (a person engaged in a struggle or contest), that assists in moving research forward by asking questions that her predecessors hoped someone would answer. Her groundbreaking work in the United States was made when she and her team proved conclusively that cannabinoid receptors do exist.

Part of a consortium of chemists, one of her findings through research at the major pharmaceutical giant, Pfizer, examined G-protein-coupled receptors of several novel potent cannabinoids. Howlett's circle found that psychotropic cannabinoids have a common ability to inhibit enzymatic function of adenylate cyclase by acting through G-proteins. Another major advancement she uncovered was made by labelling a Pfizer cannabinoid, CP55940, with radioactive tritium. This labeled species has a much higher affinity for cannabinoid receptors than the radioactive THC, which was also found to inhibit the same enzymatic function of adenylate cyclase. The results provided evidence for the presence of binding sites for this radio-cannabinoid in rat brain membranes. Those investigations yielded promising evidence that cannabinoids acted on a receptor that was G-protein coupled.

The agonist chemist was not done yet. More breakthroughs followed for CBD in 1988 when Howlett and William Devane discovered the endocannabinoid system

(ECS), which include three core components: endocannabinoids, receptors, and enzymes. The existence of receptors scattered throughout the human body had been proven to interact with substances found in cannabis, named CB1 (primarily found in the central nervous system) and CB2 (primarily found in your peripheral nervous system, especially immune cells). After the discovery of the system, it was subjected to more thorough research. It turned out that ECS is responsible for many processes that maintain balance in the human body. Because of Howlett's probing and the work of other chemists before and after, governments began to look more favorably at hemp and its components, including CBD, which they believe does not bind to CB1 or CB2 receptors like THC. Still, as she states below in an oral history interview with the author, there is still much to be accomplished [33].

In Her Own Words: Interview with Dr. Allyn Howlett, Wake Forest University:

(1) **How heavily were you impacted by the chemists/researchers that came before you when it pertains to cannabis research?**

When I began research in the cannabinoid field, it was to test the hypothesis that was proposed by Ross Johnson and Lawrence Melvin, chemists from Pfizer Central Research, that "Pfizer's analgesic compounds based upon the structure of 9-nor-9-OH-THC were promoting pain relief by acting as antagonists to prostaglandin receptors." At the time, I was working with cloned neuroblastoma cells that responded to prostaglandins with specific cellular signaling. When I tested the Pfizer compounds and THC in my assays, I found that the results did not support the hypothesis of acting as competitive blockers of prostaglandins.

So, I proposed that the compounds were working via new G proteins that we now know of as Gi proteins. At the time, I undertook an extensive search of the research literature and was astounded to find out that there was no clear understanding of how cannabinoid compounds produced their responses. Not being from that field, I made an effort to learn from the publications of researchers who had come before me about the pharmacokinetic properties of cannabinoids, how they are metabolized and what the metabolites activities were in animal models, the physical properties including how cannabinoid compounds stick to glass, plastic and polyethylene tubing and how to try to keep the compounds from sticking. All of this information was published, including information about how the both psychoactive and non-psychoactive compounds are absorbed by cell membranes and alter membrane fluidity, thereby perturbing certain cell functions when present in high concentrations. If I had not learned all of this from the research literature, I would not have been able to develop the assays and propose the hypotheses to move forward on the studies that I wanted to do.

I think that your question is very important because most reports of high-impact research findings make it appear that a single researcher suddenly had a notion that led to their discovery. The lay readers never grasp the reality that no single researcher can develop and test a hypothesis without a considerable reading of insights others have determined before them.

(2) **If you could name one aspect of your own research into cannabis-related topics that you are most proud of, what would it be?**

Many other researchers studying cannabinoid actions have commented to me that my discovery and characterization of a receptor for THC and its psychoactive and analgesic analogs "legitimatized" the field of cannabinoid research. Before our discovery that there was a receptor in the brain for psychoactive cannabinoid compounds, studies of cannabis compounds on animal behaviors was not as well-respected because it was felt (by some) that these responses were "non-specific" effects of changing membrane fluidity or some other physical property of highly lipid-soluble compounds.

The other aspect of discovering the cannabinoid receptor in the brain was that my radioligand binding and cellular signaling assays were adopted by laboratories all over the world to investigate many aspects of brain functions altered by active THC analogs that work through the cannabinoid receptors.

(3) **Where do you see CBD research going over the next 20 years in the US and elsewhere?**

Logically, CBD must exert some important mechanism(s) of neuronal regulation in the brain because we see that CBD is an effective anti-convulsant agent. However, it is not clear how CBD works in terms of cellular signaling. Although, like many other useful medicines, the therapeutic use does not require that we know the cellular mechanism(s). However, if we could pinpoint the cellular action(s), we could use that information to identify other processes in the brain that could be targeted for pharmaco-therapeutic development of CBD analogs or blockers of CBD. Some examples of current unmet therapeutic needs that might involve CBD are schizophrenia and inflammatory diseases including neurodegenerative diseases.

(4) **As a woman, can you talk about the challenges you faced in building your career in research? Were there moments when you felt that men in the field discriminated against you because of your gender?**

We all have priorities that we set for ourselves based upon our values and interests in addition to our skills. One aspect of my life that I value highly is motherhood. My career progression was based upon assessing my goals and choosing opportunities accordingly. My scientific development through my Ph. D., postdoc, and assistant professorship were of high priority during my twenties and early thirties. My daughter was born when I was mid-thirties after I had already developed an active laboratory of trainees, established my teaching, and achieved tenure, so I had opportunity to accommodate my enhanced family lifestyle. I have always felt that my male colleagues respected my skills and accomplishments. I believe that the organizational skills, ability to focus on high priority issues, and leadership that propelled my success as an academic faculty member and researcher, were all skills that I attained from my mother's upbringing, not from male mentors during my scientific training [34].[22]

3.2.4 Heir 3: Research in the Void, the Antagonist Chemist

Dr. Allyn Howlett, as a veritable agonist, achieved a career arc that represents how collaborative research on cannabis and its receptors can move forward. She worked assiduously at a time when federal regulation in America was getting stronger. To illustrate what the heirs to Roger Adams' research have grappled with, we need not look further than the pivot point beginning in the 1960s. Under the International Single Convention on Narcotic Drugs (1961) and the Controlled Substances Act (1970), the federal government became the sole "agent" for all marijuana research —both clearinghouse and gatekeeper. As of now, the DEA has only authorized one grower, the University of Mississippi (UM), which conducts their research under contract with the National Institute on Drug Abuse. As part of the deal, UM holds a DEA Schedule-I Bulk Manufacturer registration to cultivate plants for this purpose. By 2016, the DEA reversed past precedent by announcing a new interpretation of the Single Convention to allow other growers to cultivate marijuana to supply researchers. Yet, as of July 2019, there are no other laboratory initiatives approved under this new program [8].

Examining this chemical project, we find a solo laboratory, a long-game effort, by the University of Mississippi's School of Pharmacy's National Center for Natural Products Research (NCNPR). Their mission has been to move towards a comprehensive approach to natural products research. As interest in marijuana research had grown throughout the late'60s, and the government wanted to investigate the effects of cannabis on humans. This was ironic considering its past history from the 1920s on. As they reasoned it, they required a source of certifiable plant material that was responsibly grown and harvested. Coy Waller, who would later become a driving force of behind UM's operation, served on a committee of the National Institute of Mental Health and recommended that a center be created to provide marijuana research to the government. After a "supposed" open competition among institutions around the country, UM won the first contract in spring 1968. It was a prophetic moment. Later that year, UM researchers grew the first crop of legal research marijuana in the country. The "controlled" marijuana field at UM has laid fallow since 2014, but the National Institute on Drug Abuse (NIDA)

---

[22]Dr. Allyn Howlett is a biochemical neuropharmacologist, whose expertise is on cannabinoid receptor signal transduction. She is a Professor of Physiology and Pharmacology, School of Medicine at Wake Forrest University in Winston-Salem, North Carolina. The author would like to thank her for the interview. None of her responses were altered or edited in any way.

has requested a larger crop to be produced in 2019, so university officials are readying the field for use [35].

The University of Mississippi's longevity is a result of decades of policy revision at the hands of the NIDA. The operation has weathered tough competition for the contract, for which UM competes every three to five years. The Program's leadership over the past forty years continues to be fostered by the adept Dr. Mahmoud ElSohly of the School of Pharmacy. Under his directorship, the Marijuana Project has flourished in the sense that the project continues to be registered with the FDA as a drug manufacturer; this translates that everything they do must comply with the FDA's Good Manufacturing Practices. UM's Marijuana Project is part of the National Center for Natural Products Research. Ikhlas Khan, NCNPR director, continues to be affiliated with the project. As he sees it, "We are very proud of our contribution towards the science and understanding of cannabis chemistry, pharmacology and product development over last 50 years." The logic behind such an argument focuses on the idea that since the university produces marijuana based on what researchers request from NIDA's Drug Supply Program; thus, they feel they are contributing mightily to a variety of chemical projects and products. Recently, the Project has begun creating marijuana products for research, such as extracts containing THC and cannabidiol [8].

Herein lies this issue. The University of Mississippi is a chemical void. Here is why. Since they chose to partner with the largest controller in the world of marijuana, the United States government, they are subject to their will. A case in point. During the 2019 growing season, NIDA has exercised the option in its contract to grow marijuana in about half the university's marijuana field, which consists of around 12 acres of land—a fortress of marijuana, equipped with large fences and heavy security. For the past several years, the project was able to fulfill NIDA's demand for smaller amounts of material by growing in its 1,100-square-foot indoor room. The exact amount for cultivation is not yet agreed upon. Clearly, they are interested in the festooned CBD market, but it is unclear at this point how the raw material will be tested and apportioned [36].

The spark for a such a decision will be incredibly impactful on the open market. The reason behind this move could be a clinical study on CBD's impact on certain forms of juvenile epilepsy began at the University of Mississippi Medical Center. This has garnered national and international attention as a possible "cure" for what is a fatal disease. Utilizing standardized CBD extract produced on campus, the University is also conducting research on new pharmaceutical dosages in several forms that contain cannabinoids. ElSohly and his team are incredibly guarded about the Project's activities. Although they appear to be good colleagues (ElSohly has published widely on the growing strength of THC-based marijuana and its potency), behind-the-scenes they are incredibly hampered by the arrangement that was struck. Afterall, Schedule I status for the drug has its advantages, and UM remains the only place in the country that produces legal marijuana grown within federal requirements and standardized for research. About 4 years ago, the DEA announced that it would approve registrations for other qualified growers to produce and distribute marijuana for research purposes, but this has not come to fruition [37]. As for UM and ElSohly, they remain a chemical antagonist, all the while holding keys to testing and the secrets of dosage that have yet to be unlocked for CBD. An unknown chemical heritage for the future has yet to be named.

References

1. Adams' Publications, When drugs mold the future, watch out for dictators, Roger Adams Papers, University of Illinois Archives
2. Zachary GP (1999) Endless frontier: Vannevar Bush, engineer of the American century. MIT Press, Boston
3. Sullivan N (2016) The Prometheus bomb: the Manhattan Project and government in the dark. University of Nebraska Press, Lincoln, NE
4. Adams' Mailing Lists for Publication, V. Bush, defining the research task in government, Roger Adams Papers, University of Illinois Archives
5. Doel R (2004) Roger Adams: linking university science with policy on the world stage. In: Hoddeson L (ed) No boundaries, Univ Illinois Press, Champaign-Urbana
6. Decorte T, Potter G, Bouchard M (eds) (2016) World wide weed: global trends in cannabis cultivation and its control. Routledge, New York
7. Hawley E (1966) The New Deal and the problem of monopoly: a study in economic ambivalence. Princeton University Press, Princeton
8. Mosher C, Akins S (2019) In the weeds: demonization, legalization, and the evolution of U.S. marijuana policy. Temple University Press, Philadelphia
9. Lee M (2012) Smoke signals: a social history of marijuana—medical, recreational and scientific. Scribner, New York
10. Mosher C (2011) Anslinger, Harry. In: Kleiman M, Hawdon J (eds) Encyclopedia of drug policy. Sage, Thousand Oaks
11. Abel E (1980) Marijuana: the first twelve thousand years. Premium Press, New York
12. The Laguardia Committee Report New York, USA (1944) The marihuana problem in the city of New York mayor's committee on marihuana, by the New York Academy of Medicine City of New York. https://daggacouple.co.za/wp-content/uploads/1944/04/La-Guardia-report-1944.pdf. Accessed 15 Dec 2019
13. Tarbell DS, Tarbell T (1982) Roger Adams, 1889–1971. Biogr Mem (Natl Acad Sci U S A) 53:1–47
14. Adams' Publications, The chemistry of synthetic drugs, Roger Adams Papers, University of Illinois Archives
15. Doel R (2004) Roger Adams: linking university science with policy on the world stage. In: Hoddeson L (ed) No boundaries. University Illinois Press, Champaign-Urbana
16. Adams' Speeches, What the chemist has done for the physician, Roger Adams Papers, University of Illinois Archives
17. Adams' Speeches, The chemistry of marihuana, Roger Adams Papers, University of Illinois Archives
18. Adams' Publications, Marihuana, Roger Adams Papers, University of Illinois Archives
19. Adams' Speeches, Marihuana, Roger Adams Papers, University of Illinois Archives
20. Adams' Publications, Marihuana research, Roger Adams Papers, University of Illinois Archives
21. Polizzotti M (2018) Why mistranslation matters; would history have been different if Khrushchev had used a better interpreter? N Y Times. 28 July 2018 https://www.nytimes.com/2018/07/28/opinion/sunday/why-mistranslation-matters.html. Accessed 30 Nov 2019
22. US Govt (1950) NSC 68: United States objectives and programs for national security https://fas.org/irp/offdocs/nsc-hst/nsc-68.htm. Accessed 30 Nov 2019

23. Solzhenitsyn A (1968) Cancer ward. Modern Library, New York
24. Clarfield AM (2006) Novel medicine: Cancer ward, Alexander Solzhenitsyn. J R Soc Med 99 (12):641
25. Wolfe A (2018) Freedom's Laboratory: The Cold War struggle for the soul of science. Johns Hopkins University Press, Baltimore
26. Clarke A, Mamo L, Fosket JR, Fishman J, Shim J (eds) (2010) Biomedicalization: technoscience, health, and illness in the U.S. Duke University Press, Durham
27. Klein (2015) Mechoulam: the scientist, documentary. http://mechoulamthescientist.com/. Accessed 24 May 2019
28. Mechoulam R (1986) The pharmacohistory of cannabis sativa. In: Mechoulam R (ed) Cannabinoids as therapeutic agents. CRC Press, Boca Raton, FL, pp 1–19
29. Pertwee RG (2006) Cannabinoid pharmacology: the first 66 years. Brit J Pharmacology 147 (Suppl 1):S163–S171. https://doi.org/10.1038/sj.bjp.0706406
30. Kogan NM, Mechoulam R (2007) Cannabinoids in health and disease. Dialogues Clin Neurosci 9(4):413–430
31. Mechoulam R, Hanus L (2000) A historical overview of chemical research on cannabinoids. Chem Phys Lipids 108:1–13
32. Howlett A, Barth F, Bonner TI, Cabral G, Casellas P, Devane WA (2002) International Union of Pharmacology. XXVII. Classification of cannabinoid receptors. Pharmacol Rev 54:161–202
33. Howlett A (2005) Cannabinoid receptor signaling cannabinoids. In: Pertwee RG (ed) Handbook of experimental pharmacology, vol 168. Springer, Heidelberg, pp 53–79
34. Howlett A (2020) Oral history interview. Wake Forrest University, Winston-Salem
35. Duprie SS (2018) Federally funded marijuana turns 50: university researchers observe half-century of growing contract. Univ Miss News. https://news.olemiss.edu/federally-funded-marijuana-turns-50/. Accessed 23 May 2019
36. Lambert D (ed) (2009) Cannabinoids in nature and medicine. Wiley-VCH, Weinheim
37. Mitchell T (2019) DEA announces plans to increase marijuana growing licenses for research. Westword. https://www.westword.com/marijuana/dea-announces-plans-to-increase-marijuana-growing-licenses-for-research-11461841. Accessed 20 Nov 2019

Chapter 4
Epilogue: The Future Bonds of CBD

People have been using cannabis forever. The question now is, how do we as scientists catch up?

—Staci Gruber, associate professor of psychiatry at Harvard Medical School [1].

The brain is about a symphony, and CBD can bring the entire symphony into harmony.

—Yasmin Hurd, Ph.D., Director of the Addiction Institute at the Icahn School of Medicine at Mount Sinai Hospital [2].

4.1 "Trusted" Chemists: The CBD Dispensary

At a local CBD store in Houston, Texas, a young man in his twenties offers a greeting from behind the counter, "You know I just started here," he grudgingly admits, "I have only studied this stuff for a week or so." His background was in retail, and he looked the part as if he had just stepped out of a Banana Republic advertisement. He seemed confident that he could speak with a measure of authority, as he rattled off statistics, pointed out products, and made the experience of information gathering an enjoyable experience (Fig. 4.1).

The phenomenon of the "CBD Dispensary" has taken hold—both in the form of a physical storefront and massive online presence. In even the most conservative bastions, you will find them. Some are seedy and not well-appointed; while others are highly stylized with thought-provoking displays carefully designed to draw the eye and the pocketbook. The products are limitless with lotions, coffees, pet treats, gum, and many more infusions that are possible. CBD seems to be in everything.

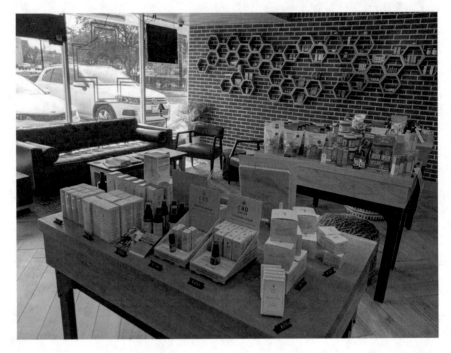

Fig. 4.1 Example of twenty-first century "CBD Dispensary" (Photo courtesy of the author)

How much knowledge these businesses have of chemical history is debatable, but they exist as an extension for a drug that ranges from misunderstood and feared to a panacea for pain [3].[1]

Time and time again throughout this project, when I told people I was working on CBD they wanted to try to understand how the dosage worked and the notion of "trust" was mentioned because they have seen the massive advertisement campaigns for the product. As this study has tracked, from Maltos-Cannabis to the present, distrust continues to be a regular theme, as consumers try to decide if investing in something that is medically difficult to track through the generation of data. Today, even after almost 80 years of research and development, we still have much to learn about CBD's properties and capabilities. What we do know is that CBD is a technically legal even in places like the United States, but the major issue it faces is that since laboratories are unable to legally perform tests on the product, there is no means to tell its effects—more data continues to be needed [4].

---

[1]In America, the FDA attempts to regulate CBD stores that sell illegally branded items, but it is difficult to keep up. CBD Dispensaries can be found throughout the world. Estimates of the cannabis industry for the future equate to billions of dollars. Interestingly, during the COVID-19 Pandemic, CBD stores were legally successful at arguing they were "essential businesses." https://www.houstonchronicle.com/politics/texas/article/Texas-shop-selling-CBD-oil-challenges-county-15172776.php.

Oversight is evolving and though we may not exactly know the impact that this compound is making on society at-large, it continues to be marketed and discussed in a variety of circles. As this study sought to prove, attempts at isolation and exploration were attempted before the Second World War. Strangely enough, confusion over products sans THC has led to a lack of uniform regulations governing this segment, thus a definition remains elusive. What is more certain is that chemistry, as a pursuit, could not have arrived at this point in history without the people that investigated, probed, and then stood on the shoulders of others—a chemical heritage.

4.2 Tied Conclusions

Charlotte Figi and Dr. Sanjay Gupta probably did as much for CBD as anyone in the world. In a documentary called *Weed*, Gupta, a renowned physician who is the Chief Medical Correspondent for CNN, spotlighted the chronic epileptic seizures of Figi. Plagued by Dravet Syndrome, a rare form of epilepsy, it could not be controlled by regular medication. At age eight, she was treated by a CBD oil produced by a local Colorado cannabis grower and the success was magnificent. Her debilitating seizures were vastly reduced, and people who saw the documentary were persuaded that isolating THC could lead to medical miracles. Popular culture embraced CBD like never before, just as jurisdictions grappled with how to make sense of it all—was CBD, marijuana? Almost every state in America has embraced "medical marijuana" for the treatment of pain. Even the most conservative governments have embraced its possibilities. Still, for CBD, political and medical lines continue to be drawn and drug makers at all levels continue to race into the marketplace [5].[2]

Impending prospects for CBD revolve around research and data, not soothsaying; in other words, chemistry belongs in the hands of chemists. When Raphael Mechoulam conducted in-depth research on cannabinoids over 50 years ago, he started a revolution by asking new questions with some older data—namely, the work of Roger Adams. To him, goes most of the credit in the popular press for "discovering" CBD and dealing with myths about cannabidiol, but he was heir to an already established legacy. During the early 2000s, main factors that supported evidence of the existence of cannabinoid receptors in the body were heavily influenced by Mechoulam, Allyn Howlett, and several others (see Appendix). They were first to uncover that the pharmacological activity of psychotropic cannabinoids was influenced by the chemical structure. Also, ideas chemically trickled down that cannabinoids with chiral centers exhibit stereoselectivity. Thus,

---

[2]Charlotte Figi's life was extended for another five years. Sadly, in April 2020 she became ill during the COVID-19 Pandemic and passed away; possibly due to the virus. News outlets mentioned in several stories that Figi was the girl who sparked the CBD Movement. https://www.cnn.com/2020/04/08/health/charlotte-figi-cbd-marijuana-dies/index.html.

the potency of THC matches that of known classes of receptors. Despite access from entities of privilege, like the University of Mississippi, these are some of the exciting challenges that marijuana testing and the exploration of CBD probe—that is the future, especially concerning the challenges of addiction [6].

Roger Adams thought deeply about what was ahead for chemistry. As an early advocate of the hybrid biochemistry, he originally began his investigations of CBD for several reasons. It included thinking about the prosecution of pain, his developing philosophy on the role of medicine, and kinship with his fellow man. He fundamentally understood that discoveries did not just happen randomly; rather, if you asked questions, thought differently about problems, and pursued scientific inquiry down untrodden paths, then you might find something revolutionary at the end, or not. Adams did not "discover" CBD per se, but those that came after him were part of a bond through time that linked chemical investigations from one generation to another. The origin and development of CBD is a complicated one. The search for the perfect food persists. As stores peddle their wares, as governments wrestle with legalities, the chemical evolution continues at a modulating pace regulated by professionals. In the end, that is the how scientists have caught up and learned to save all the parts—the bonds that tie.

References

1. Velasquez-Manoff M (2019) Can CBD really do that? NY Times Mag, 14 May 2019. https://www.nytimes.com/interactive/2019/05/14/magazine/cbd-cannabis-cure.html?mtrref=www.google.com&gwh=0522F857222CD5FABB422CFCF9744FBB&gwt=pay&assetType=REGIWALL. Accessed 23 May 2020
2. Hurd Y (2019) Can CBD Really do that?—Moises Velasquez-Manoff. https://www.mountsinai.org/about/newsroom/2019/can-cbd-really-do-all-that-moises-velasquezmanoff. Accessed 23 Sept 2020
3. FDA (2019) Warning Letters. https://www.fda.gov/inspections-compliance-enforcement-and-criminal-investigations/compliance-actions-and-activities/warning-letters. Accessed 1 Dec 2020
4. Pertwee RG (2006) Cannabinoid pharmacology: the first 66 years. Brit J Pharmacology 147(Suppl 1):S163–S171. https://doi.org/10.1038/sj.bjp.0706406
5. Orrin D, Patel AD, Helen CJ, Vicente V, Wirrell EC, Michael P, Greenwood SM, Claire R, Daniel C, van Landingham Kevan E, Zuberi SM (2018) Effect of cannabidiol on drop seizures in the Lennox-Gastaut Syndrome. N Engl J Med 378:1888–1897
6. Grisel J (2019) Never enough: the neuroscience and experience of addiction. Doubleday, New York

Appendix

The more recent history of chemical research on the cannabis sativa plant that proffered the discovery of the endocannabinoid system has the potential to expand our understanding of the capability of CBD. How we interpret these events remains to be seen, since their impact, as Dr. Allyn Howlett suggested. Here are some of the recent, important highlights. However, these are not all.[1]

The Process of Discovering the Endocannabinoid System

1990: Receptor Pinpointed in DNA

Lisa Matsuda announces that she and her colleagues have identified a DNA sequence that defines a THC-sensitive receptor in a rat's brain to the National Academy of Science's Institute of Medicine. Soon after, this receptor was successfully cloned allowing them to fashion molecules that activate the receptors. Genetically altered rats were also bred that lacked such a receptor, meaning THC would have no effect on them. Success in these experiments proved that THC works by activating specialized cannabinoid receptors found in both the brain and central nervous system, meaning they must be present in humans for effects to be felt.

---

[1]Timeline taken from the following website: https://www.hytiva.com/learn/discovering-the-endocannabinoid-system.

J. N. Campbell, *Bonds That Tie: Chemical Heritage and the Rise of Cannabis Research*, History of Chemistry, https://doi.org/10.1007/978-3-030-60023-5

1992: First Endocannabinoid Discovered

Raphael Mechoulam and National Institute of Mental Health (NIMH) researchers William Devane and Dr. Lumir Hanus discover anandamide, a naturally occurring endogenous cannabinoid found in the human body. Such an endocannabinoid was found to attach to the same receptors as THC, thus being named after the Sanskrit word for bliss. Also nicknamed "the bliss receptor," anandamide plays a role in memory, pain, depression, and appetite.

1993: Second Receptor Pinpointed in DNA

A second THC-sensitive receptor is found in the immune and nervous systems, named the CB2 receptors. They're predominantly in the heart, blood vessels, kidneys, bones, gut, spleen, reproductive organs, and lymph cells.

1995: Second Endocannabinoid Discovered

Raphael Mechoulam and colleagues again find a new endocannabinoid named 2-AG (2-arachidonoylglycerol) that attaches to both CB1 and CB2 receptors.

Index

Printed in the United Sta
By Bo